Propagation Modeling for Wireless Communications

Propagation Modeling for Wireless Communications

Indrakshi Dey

CRC Press
Taylor & Francis Group
Boca Raton London New York

CRC Press is an imprint of the
Taylor & Francis Group, an **informa** business

MATLAB® is a trademark of The MathWorks, Inc. and is used with permission. The Math-Works does not warrant the accuracy of the text or exercises in this book. This book's use or discussion of MATLAB® software or related products does not constitute endorsement or sponsorship by The MathWorks of a particular pedagogical approach or particular use of the MATLAB® software.

First edition published 2022
by CRC Press
6000 Broken Sound Parkway NW, Suite 300, Boca Raton, FL 33487-2742

and by CRC Press
4 Park Square, Milton Park, Abingdon, Oxon, OX14 4RN

CRC Press is an imprint of Taylor & Francis Group, LLC

ISBN: 978-1-032-08079-6 (hbk)
ISBN: 978-1-032-08115-1 (pbk)
ISBN: 978-1-003-21301-7 (ebk)

DOI: 10.1201/9781003213017

Publisher's note: This book has been prepared from camera-ready copy provided by the authors.

Contents

List of Figures

List of Tables

Part I

Introduction

1

Introduction

CONTENTS

This chapter starts with a brief introduction into what are wireless communication systems and networks and their applications. Basic concepts and mechanisms driving the wireless propagation are briefly introduced depending on the different media (air, water, etc.) and modes (waves, particles etc.) used for communication. Readers are next motivated toward why we need to model the propagation environment, why modeling the environment is challenging and why it is crucial to come up with a model as accurate as possible. The chapter is rounded with some historical and philosophical perspectives.

1.1 Introduction

Today we cannot really imagine ourselves without a mobile phone. What is it used for? Mobile phones are used for communicating between two people that are separated by some amount of distance. According to Merriam-Webster, telecommunication is

DOI: 10.1201/9781003213017-1

defined as communications at a distance. We generally associate mobile phones with telecommunications or wireless communications. However, it means much more. It encompasses electrical communication of voice, data, and image over a distance, for example, television, fax, satellites, and Wi-Fi. That's why Institute of Electrical and Electronics Engineers (IEEE) defines telecommunications as transmission of signals over a long distance such as by telegraphy radio or television. Telecommunications is also referred to as electrical communications or simply communications; still it does not involve everything that can be referred to as communication systems. Communication system includes all the devices that we have at home nowadays and that can carry out controlling or altering function. Think about not only the television but also the television remote, home security systems, garage door openers, remote controls for a home theater system, Wi-Fi, mobile broadband, or fiber broadband and modern. All of these are small, midsized, or big communication systems.

Let us consider the ancient system of communications by pigeon post where a letter is tied to the legs of a pigeon and let it fly from one kingdom to another. In this case, the letter will contain some message like "there is a flood, need some help" or "there is a war" or something similar. This message can be referred to as information. The letter is the format in which the information is communicated. Let's refer to the letter as the signal. The pigeon is the one that carries the letter. Let's refer to it as the medium. So basically, a communication system consists of three important elements: information, signal, and medium. Therefore, communication deals with conveying information using signals over a medium. Communication systems can be either wired or wireless. If there is a physical connection between the transmitter and the receiver, the system is referred to as wired communication system while if there is no actual physical connection, the system is referred to as wireless communication system. Throughout this book, we will consider a wireless communication system.

1.2 The Propagation Environment

There are few things in nature more unwieldy than the power-limited, space-varying, time-varying, frequency-varying wireless channel signals propagating through the wireless environments. The wireless environment suffers from small-scale fading (caused by transmission via different reflectors resulting in a large variety of propagation path delays), shadowing (caused by obstacles of size much larger than the propagating wavelength), scattering (caused by the interaction of the traveling waves or any modes of communication with objects of dimensions on the order of wavelength or less), diffraction (caused by bending of traveling waves around an obstacle and the charge change impact loss due to the variation in relative distance between the transmitter and the receiver).

The wireless propagation environment is comprised of different kinds of terrains, trees, buildings, vehicles, and people that diffract, scatter, and absorb the wireless signal that travels through the path between the transmitter and the receiver. These obstacles along the signal path result in signal attenuation greater than what is experienced in free space. Copies of the actual signal travel over multiple paths between the transmitter and the receiver as the signal is scattered by obstacles. This will result in time and angular dispersion of the signal as multiple copies arrive with different delays and from different directions, respectively. Fast variations in the phase relationship between these multiple signal copies result from moving scatterers and or either of the transmitter and receiver resulting in deep fades and the so-called frequency dispersion. Antenna type and configuration with the range of their incident angles dictates the correlation between fading observed at the output of the receive antennas.

1.3   Need for Propagation Modeling

The main aim of channel modeling is to assimilate knowledge and understanding the way in which the propagation environment affects, impairs, and distorts the troubling signals into a format such that they can be easily used for design test and simulation of wireless communication systems. Communication system designers, engineers, and developers use these channel models to estimate, predict, evaluate, and compare performances of different communication systems and in turn propose, develop, and validate different techniques for improving performance and mitigating impairments. Modeling the propagation environment forms the fundamental basis for developing system software, channel emulators, and planning tools, that in turn are used to design, develop, implement, and deploy a complete wireless communication system.

A channel model is an attempt to represent the practical propagation environment through simple but abstract mathematical framework. Since channel models are supposed to replicate the real environment as closely as possible but with low complexity, it has become imperative to develop new channel models as researchers come up with novel applications and novel communication systems using different media and carriers of information. With systems being deployed in ever more challenging environments, it has also become necessary to extend existing channel models.

1.4   Historical Highlights

Since the earliest days of the advent of wireless systems, the need to understand and characterize the propagation environment was paramount effect of iron spear fire on radio wave propagation that was first started during the 1920s and 1930s [1]. These studies introduced the conceptual assumption of WSSUS channels, which is still used today. Effects of troposphere hydrometers and ground reflectors were studied during the Second World War while developing radars [2]. Results from these studies were used to design and deploy point-to-point microwave systems, land mobile, and cellular systems. Separate key modeling efforts in the 1960s and 1970s set the stage for planning of modern radio systems including different models for path loss calculation [3], Clark's model [4] for Doppler spectrum characterization of mobile radio propagation scenario, Bello's model [5] for analyzing linear time-varying wide band channels and Cox's [6] model for characterizing time dispersion of white band mobile channels that is introduction of the concept of power-delay profile. 1980s give way two more well-developed full-fledged channel models like the European COST 207 [7] action that provided the foundation for development of GSM the European standard for 2G telephony [8].

The success of COST 207 set the stage for follow-on ventures, including COST 231 – Digital Mobile Radio Towards Future Generation Systems (1989–1996), COST 259 – Wireless Flexible Personalized Communications (1996–2000), COST 273 – Towards Mobile Broadband Multimedia Networks (2001–2005), and COST 2100 – Pervasive Mobile and Ambient Wireless Communications (2007–2009). After the COST, more collaborative task forces have been set up like IEEE 802.16 Working Group on Broadband Wireless Access Standards, IEEE 802.11's Task Group (TG) n, IEEE 802.15's TG3a and 4a, IEEE 802.15's TG3c to develop enhanced propagation models [8, 9, 10, 11].

Channel modeling for any wireless communication system starts with enlisting and narrowing down the aspects of propagation environment that are to be considered for the system design at hand. In any wireless network the mode of communications is conditioned on the medium of exchange. Electromagnetic (EM) waves (light, radio, etc.) can carry information reliably over air and free space. Unfortunately, EM waves are rapidly attenuated when traveling through water or solid earth. Acoustic waves are therefore used to carry information underwater and underground. Utilizing EM communications is challenging for emerging nano-scale applications like smart drug delivery, detection of toxic substances, or allergens, early detection of cancer cells within the human body, owing to size constraint and high attenuation caused by different body-fluids acting as the medium. In these scenarios, molecules or lipid vesicles are used as information carriers. The molecules are released by the transmitter into an aqueous or gaseous medium, where they propagate by diffusion.

Different modeling approaches are used in different kinds of system depending on the medium and the mode. Therefore, the first step is to specify the communication scenario media and mode and the impairments encountered in that scenario which are to be compensated for. The propagation phenomenon for any environment is affected by environmental factors as well as the properties of the transmit and receive units themselves. For example, properties like the gain, bandwidths, polarization, and orientations of the transmit and receive antennas affect the traveling signal. While environmental design parameters like the carrier frequency, distance between the transmit and receive antennas, relative heights of the antennas from the ground level, the nature, height and density of the scatterers, and the nature of the obstructions that lie between the transmitter and the receiver affects the propagation signal.

Modeling efforts can be broadly categorized into two groups measurement-based and simulation-based. Measurement-based methods can provide results that can be immediately used by designers and developers and validate simulated performances of the proposed system design. Measurement-based and simulation-based approaches are complementary. Simulation-based methods refer to ray tracing that can use geographical and morphological information to map the propagation environment. However, such models are very site-specific. Again, measurement campaign in turn tests and validates the simulation model for any environment.

1.5 Additional Preliminaries – Measurement of Wireless Channels

Several different measurement techniques for EM wave propagation have been used over the years depending on the communication scenario parameters and system details, some commonly used techniques are summarized below. A channel sounder can be defined as a device used for parameter estimation associated with channel impulse response (CIR), namely, the number of multipath components and their respective amplitude, phases, and path delays. A simple channel sounder transmits repetitive pulses of width T_w secs and receives the transmitted wave using a wide bandpass filter, as shown in Fig. 1.5. The received signal is then amplified, detected using an envelope detector, accumulated, and stored. The minimum resolvable delay between multipath components is equal to the probing pulse width T_w, while the repetition rate T_r determines the maximum unambiguous excess delay that can be resolved. Channel sounders have been used in several measurement campaigns for single-input-single-output (SISO), single-input-multiple-output (SIMO), and multiple-input-multiple-output (MIMO) indoor radio channel propagation scenarios. Several combinations of transmitters, receivers, switching equipment, and antenna arrays have been used to establish different measurement scenarios, a few of which is described briefly in the following subsections.

1.5.1 Indoor Wide-band SIMO Channel Sounder

A measurement campaign for SIMO indoor radio communication has been presented in [12]. The measurement set-up consists of a single transmit and two receive antennas. The receiver set-up is accomplished by combining a switching mechanism with a 1.9 Gigahertz (GHz) SISO channel sounder. The switch is characterized by means of a Vector Network Analyzer (VNA). The switching rate is larger than the Doppler shift in static environments. The channel sounder determines the complex envelope of the CIR by means of a sliding correlation technique. A schematic of the

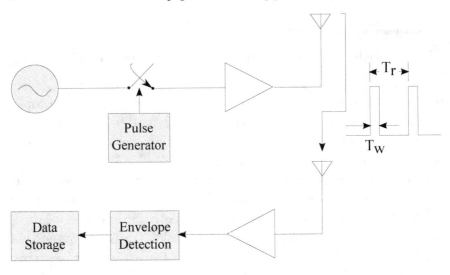

FIGURE 1.1
Block diagram of a simple channel sounder.

set-up is given by Fig. 3.1. The measurement campaigns, however, reveal a slight phase shift of the received signal as a function of time, caused by a progressive de-pairing of the oscillating crystals in the transmitter and the receiver. Hence, only the relative phase of the CIRs can be measured, while the absolute phase values and the central complex correlations cannot be measured or estimated accurately.

1.5.2 Vector Radio Channel Sounder

Designing a vector radio channel sounder, called RUSK ATM, has been presented in [13], operating in the 5 X 6 GHz HIPERLAN frequency band. It consists of a mobile transmitter excited by periodic multi-frequency wave and a fixed receiver. The time period of the excitation signal is chosen corresponding to the expected maximum channel excess delay. The transmit signal generated is upconverted to the RF range from its respective baseband and is radiated from an omni-directional monopole antenna. In case of synchronized Fast Fourier Transform (FFT) processing, leakage

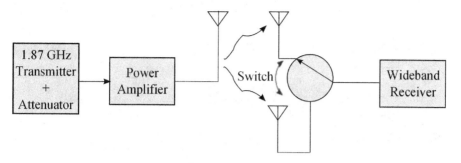

FIGURE 1.2
Measurement set-up for a SIMO Channel Sounder.

bias is effectively eliminated. The multi-frequency excitation signal comprises quadratic phases for crest factor minimization. The transmit signal spectrum is band-limited for reduction of electromagnetic interference during free field measurement. The receiver block structure is given by Fig. 1.3.

By fast multiplexing, the elements of the channel response vector snapshot (CRVS) are sequentially estimated from consecutive periods of the transmitted signal. Since the excess delay is limited because of the dynamic range and path loss, the maximum channel spread factor is well below 1%. Virtual alignment of the CRVS vector in time is achieved by an interpolation procedure, since a phase offset proportional to the Doppler shift is introduced by sequential data acquisition. The delay-domain estimate of the dominant output path weights is improved by respective correlation gain as the excitation signal is pseudo-random with a constant power spectrum. Received signal averaging reduces SNR and in turn also reduces the resolvable Doppler bandwidth.

1.5.3 Wideband 60-GHz Channel Sounder

A 60-GHz wideband channel sounder has been developed in [14] for vector analysis of the channel transfer function (CTF) by frequency transposition. The experimental set-up is exhibited in Fig. 1.4. Two heterodyne emission and reception heads have been

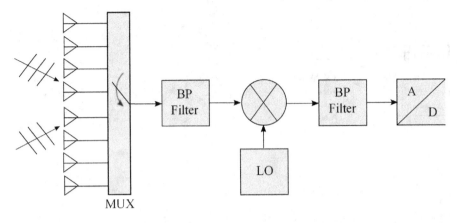

FIGURE 1.3
Receiver Structure of Vector Radio Channel Sounder (BP: Band-pass; LO: Local Oscillator; A/D: Analog-to-Digital Converter; MUX: Multiplexer).

developed by monolithic integration with frequencies ranging from 57 to 59 GHz with intermediate frequencies of 1 to 3 GHz. A VNA is first calibrated and is then used to measure the frequency transfer function. The CIR is then derived by applying an Inverse Fourier Transform (IFT). Temporal resolution is generally chosen to be 1 ns for an 800 ns impulse response (IR). To characterize the propagation phenomena, a system to accurately position the receiving antenna has been constructed in [15]. Correct measurement can be accumulated at every separation of 1 mm on a total distance of 50 cm.

1.6　Philosophical Perspective

Information is one of our most valuable resources, with humans generating an average of 2.5 quintillion bytes of information per day [16]. This figure is expected to grow significantly, as we continue to intensify the development of systems that take information about us, or the conditions around us, and make the surrounding world react accordingly. Such systems fuse our physical world with

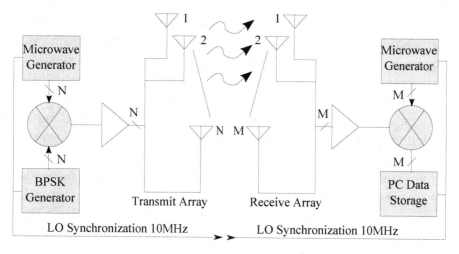

FIGURE 1.4
Measurement Set-up for 60 GHz Channel Characterization.

the digital world using the power of two "C"s; Computing and Communications. For example, wearable and implantable sensors collect data from a person and communicate them over air to the cloud. Within the cloud, advanced computing techniques are used for immediate diagnosis, real-time risk analysis, and activating appropriate action, if necessary, which are then communicated back to the relevant actors or endpoints [17]. This system should be able to communicate information irrespective of the location (indoor, outdoor, in-flight, underground) and condition (mobile, stationary) of the individuals and the actors. The above-mentioned system works by communicating information wirelessly over different media using different modes, existing infrastructure (4G/5G, IoT networks) and novel ones (nano-networks using molecular diffusion); a wireless communication system in its widest sense. Implantable sensors will communicate information over body fluids (medium) using molecules (mode), wearable and environmental sensors communicate over air or body surface (medium) using acoustic or electromagnetic (EM) waves (mode), and information propagates over air, water, and solid earth to a central platform using electromagnetic and/or acoustic waves (for individuals located underground or underwater) [18].

The wireless environment represents a diverse, intricate world, and we don't really know what happens to our information (signal) once it leaves the source. So, we try to model (theoretical, experimental, simulation-based) the environment through which the information propagates to estimate the flow of information and the signal characteristics at any point in space. However, different modeling approaches are used to describe flow of information in different kinds of network [19] depending on the medium and the mode. Traditionally, three different techniques are used to model the wireless environment: a) Deterministic (ray tracing) [20] techniques model the propagation channel in a specific location using the geographical and morphological information from a database, b) stochastic techniques [21] use stochastic distribution of scatterers, and c) geometry-based stochastic modeling techniques [22] combine ray tracing and stochastic modeling. The major problem with all these modeling techniques is that either they are location-specific or they are media and mode-specific. Therefore, such approaches fail to estimate what happens when the signal propagates from one medium to another while changing the mode of communications. This drives the need for a generalized channel model irrespective of the media and mode of information transfer.

This book makes the first-ever attempt to introduce and describe theoretical frameworks and simulation models for characterizing flow of information in wireless networks communicating over different media (air, water, underground, space, human body) of propagation using different modes (radiowave, light waves, quantum particles, molecules) as information carriers. In case of each type of network, the book uses basic concepts of physics, mathematics, geometry, and probability theory to study the impact of the dimension and shape of the propagation environment and relative transmit-receive position on the information flow. Therefore, this book lays the foundation for modeling information flow wirelessly and developing a mathematical framework capable of representing the information flow over any medium; and a corresponding simulator capable of estimating the characteristics of the transmitted signal at any point in space, irrespective of geographical location and resources used.

Bibliography

[1] A. H. Waynick, "The early history of ionospheric investigations in the United States," *Phil. Trans. R. Soc. Lond. A.*, vol. 280, no. 1293, pp. 11–25, 23 Oct. 1975.

[2] D. E. Kerr, *Propagation of Short Radio Waves*. vol. 13 of the MIT Radiation Laboratory Series. New York: McGraw-Hill, 1951.

[3] Y. Okumura et al., "Field strength and its variability in VHF and UHF land-mobile radio service." *Rev. Elec. Commun. Lab.*, vol. 16, no. 9-10, pp. 825–873, 1968.

[4] R. H. Clarke, "A statistical theory of mobile radio reception," *Bell Sys. Tech. J.*, vol. 47, pp. 957–1000, Jul.–Aug. 1968.

[5] P. A. Bello, "Characterization of randomly time-variant linear channels," *IEEE Trans. Commun. Syst.*, vol. 11, no. 4, pp. 360–393, Dec. 1963.

[6] D. C. Cox, "Delay Doppler characteristics of multipath propagation at 910 MHz in a suburban mobile radio environment," *IEEE Trans. Antennas Propag.*, vol. 20, no. 5, pp. 625-635, Sep. 1972.

[7] Jing Jiang et al., "Energy Efficiency and Optimal Power Allocation in Virtual-MIMO Systems," *IEEE Vehicular Technology Conference (VTC Fall)*, 6 pages, Sept. 3-6, 2012.

[8] W. Jakes, Ed., *Microwave Mobile Communications*, New York: Wiley, 1974.

[9] D. Greenwood and L. Hanzo, "Characterization of mobile radio channels," in *Mobile Radio Communications*, R. Steele, Ed., London: Pentech Press, pp. 92–185, 1992.

[10] H. L. Bertoni, W. Honcharenko, L. R. Maciel and H. H. Xia, "UHF propagation prediction for wireless personal communication," *Proc. IEEE*, vol. 82, no 9, pp. 1333–1359, Sep. 1994.

[11] A. F. Molisch, *Wireless Communications*. New York: Wiley, 2005, pp. 43–170.

[12] L. P. Rice, "Radio transmission into buildings at 35 and 150 mc," in *Bell Syst. Tech. J.*, vol. 38, no. 1, pp. 197-210, Jan. 1959.

[13] H. H. Hoffman and D. C. Cox, "Attenuation of 900 MHz radio waves propagating into a metal building," in *IEEE Trans. Antennas Propagat.*, vol. AP-30, no. 4, pp. 808-811, July 1982.

[14] D. C. Cox, R. R. Murray, and A. W. Noms, "Measurements of 800 MHz radio transmission into buildings with metallic walls," in *Bell Syst. Tech. J.*, vol. 62, 110. 9, pp. 2695-2717, Nov. 1983.

[15] D. C. Cox, R. R. Murray, A. W. Noms, "800-MHz attenuation measured in and around suburban houses," in *ATT Bell Lab. Tech. J.*, vol. 63, no. 6, pp. 921-954, July-Aug. 1984.

[16] Url: https://blog.microfocus.com/how-much-data-is-created-on-the-internet-each-day/.

[17] S. Siddiqui, A. A. Khan, F. Nait-Abdesselam and I. Dey, "Anxiety and depression management for elderly using Internet of Things and symphonic melodies," Accepted to *2021 IEEE International Conference on Communications (ICC): SAC E-Health Track - E-Health*, Montreal, Canada.

[18] I. Dey, G. G. Messier and S. Magierowski, "Joint Fading and Shadowing Model for Large Office Indoor WLAN Environments," *IEEE Transaction on Antennas and Propagation*, vol. 62, no. 4, pp. 2209–2222, Apr. 2014.

[19] I. Dey and N. Marchetti, "A Simulink-based Channel Emulator for Underwater Acoustic Communications", *IEEE TechRxiv*, https: // www.techrxiv.org.

[20] A. Hsiao, C. Yang, T. Wang, I. Lin and W. Liao, "Ray tracing simulations for millimeter wave propagation in 5G wireless communications," *2017 IEEE International Symposium on Antennas and Propagation & USNC/URSI National Radio Science Meeting*, 2017, pp. 1901-1902.

[21] I. Dey, P. Salvo Rossi, M. M. Butt and N. Marchetti, "Experimental Analysis of Wideband Spectrum Sensing Networks using Massive MIMO Testbed," *IEEE Transactions on Communications*, vol. 68, no. 9, pp. 5390-5405, May 2020.

[22] A. Roivainen et al., "Parametrization and Validation of Geometry-Based Stochastic Channel Model for Urban Small Cells at 10 GHz," in *IEEE Transactions on Antennas and Propagation*, vol. 65, no. 7, pp. 3809-3814, July 2017.

Part II

Wave-based Propagation

2

Radiowave Propagation

CONTENTS

DOI: 10.1201/9781003213017-2

The flow of signal (information and energy) through any medium is guided by the characteristics of the mode of propagation. The most common mode of communication information wirelessly is electromagnetic (EM) waves. EM waves can be visible (light) or invisible (radio) depending on their range of frequency. The speed of EM waves in free-space corresponds to the speed of light and is equal to 3×10^8 m/sec. The world of electromagnetic is guided by Maxwell's equations, a symmetric coherent set of four complicated equations. The equations are rules the universe uses to govern the behavior of electric and magnetic fields. A flow of electric current will produce a magnetic field. If the current flow varies with time (as in case of any periodic signal), the magnetic field will also give rise to an electric field. Maxwell's equations also show that separated charge (positive and negative) gives rise to an electric field and if this is varying in time as well, will give rise to a propagating electric field, further giving rise to a propagating magnetic field. In EM wave-based communications, propagating waves interact with the environment and gets reflected, diffracted, scattered and refracted by structures within the environment. As a result, both amplitude and phase of the transmit signal fluctuate over time causing spatio-temporal variations within the signal itself. This chapter will provide a detailed overview of different models that have been proposed over the years for characterizing the flow of EM waves and their interaction with the propagating environment depending on the medium of information exchange.

2.1 Introduction

The notion of electromagnetic (EM) waves propagating with constant speed in homogeneous media was introduced by James Clerk Maxwell in 1865, based on relations between varying electric and magnetic fields. The speed of EM waves in free space corresponds to the speed of light and is equal to 3×10^8 m/s. Radio waves have a nature similar to EM waves was invented by German scientist named Hertz; however they are invisible.

Receive antenna collect radio waves radiated from the transmit antenna by receiver antenna after propagating between antennas. Several phenomena affect them during propagation. The imposed effects are due to transmission media as a function of radiowave characteristics such as frequency bandwidth, polarization, and type of signal.

In general, radiowaves are some type of electromagnetic waves in a specific frequency band. Although a distinctive spectrum is not determined for radiowaves, in this book propagational phenomena are limited to frequency bands devised by International Telecommunications Union (ITU), i.e., 3 KHz to 275 GHz. Due to the key role of radio communications, a lot of studies, researches, and efforts have been spared by competent experts, institutes, organizations, and administrations all over the world for radiowave propagation and using higher frequency bands. Major results and achievements in radiowave propagation are collected and summarized in series P of ITU-R recommendations, which are revised and updated periodically.

2.2 Basic Preliminaries and Details

Radiowaves are used for transmitting various kinds of audio, video, data, control, and navigational signals, encompassing ariel, land,

and maritime mobile services, navigational aids, satellite communication services, audio and video broadcasting, telemetry, remote
sensing, traffic control systems, radio special services for indus-
trial, scientific, research, medical, and social applications.

The propagation channel behaves differently at different frequencies of operation depending on the geographical environment.
With the exception of Asia, the ITU has allocated 600 MHz bandwidth at 48/47 GHz (shared with satellites) worldwide for HAP
services. The 31/28 GHz band has been selected for Asia since
48/47 GHz band is susceptible to rain attenuation, creating a serious problem in Asia and tropical regions [14]. Another option is
the W band (70–80 GHz) with up to 5 GHz of bandwidth available world-wide for up and down link. For example, the W band
can be a promising solution for connectivity between high mobility aeronautical terminals and the high altitude platforms (HAPs).
Therefore, the choice of frequency band for both information transfer and wireless power transfer will depend on the geographical
location of the transceivers.

Radio systems can be classified based on the frequency bands
of operation, like Low Frequency (LF), Medium Frequency (MF),
High Frequency (HF), Very High Frequency (VHF), Ultra High
Frequency (UHF), Super High Frequency (SHF), and Extra High
Frequency (EHF) radio systems. ELF, ULF, and VLF have very
limited use, owing to the requirement of very big antennas and
poor propagation characteristics, low bandwidth availability resulting in very low data rate for transmission, and are used for sub-
marine telegraphy communications and high atmospheric noise.
LF (30–300 KHz) can be used as ground waves in short-distance
communications, ground-based waveguide for long-distance communications for broadcasting time signals, radio navigational aids;
however, they suffer from high atmospheric noise and limited
bandwidth. MF (300–3000 KHz) can be used as ground waves in
short-distance communications, ground-based or ionospheric hops
in long-distance communications specially during nights for radio
broadcasting services and maritime mobile and radio navigation
services.

HF band (3–30 MHz) is used in Ionospheric long haul hops for radio broadcasting services, aeronautical and maritime mobile communications. VHF Band (30–300 MHz) is used in long distance receiving for audio and video broadcasting, aeronautical and maritime radio communications, over-horizon radio communication by tropo-scatters, radar and radio navigation services, analog cordless telephone, radio paging services and lower earth orbit (LEO) satellite systems, while UHF band (300–3,000 MHz) is used for TV broadcasting, cellular mobile radio services, mobile satellite, GPS, astronomy communications, radar and radio navigation services. SHF band (3–30 GHz) is used for radar systems and military applications, TV satellite broadcasting and remote sensing from satellites. EHF band (30–300 GHz) is being planned for broadband-fixed wireless access, future satellite and high-altitude platform applications. Similarly, micro-metric and nano-metric bands are being planned for space radio communications, satellite communications, laser and infrared radio communications, and fiber optics cable networks.

Based on the application range and scenario, radiowave communication systems can be classified as aeronautical or maritime radio navigational aids, audio or video broadcasting, fixed- or mobile-satellite services, aeronautical, land, and maritime mobile services, Point-to-point or point-to-multipoint fixed radio services, radar systems, wireless communications, radio special services and radio amateur services. Each of the above services usually includes several systems. For example, a satellite radio system may be referred to by each one of the items like fixed systems like intelsat and eutelsat, mobile systems like inmarsat, meteorological satellite such as meteosat, earth exploration systems, remote sensing and telemetry, audio and video broadcasting, navigational aids and radio determination systems, military systems, national and regional satellite systems such as vsat, insat, arabsat, and gsat.

2.3 Propagation Phenomenon

Formulating a realistic radio channel model for wireless propagation faces several challenges and design issues, due to the random and time-varying nature of the communication channel. In a wireless communication system, the transmitted signal disperses in almost all directions possible. Reliable wireless communication can only be ensured through the provision of adequate signal level at the receiver. Knowledge of the impact of the wireless channel inconsistencies on the transmitted signal is essential for the design of a reliable communication system. Detailed characterization of these channel impairments must be achieved before efficient communication ensues. In the following, we summarize some of the factors responsible for degrading the quality of the received signal, and hence are needed to be investigated for successful design of a communication system.

2.3.1 Spatio-time Variation of the Channel

Radiowave communication through wireless channels is a very complicated phenomenon. Propagating waves suffer from reflection, diffraction, scattering, and refraction by structures inside a building. These phenomena are broadly grouped as fading and shadowing. When the transmitted signal experiences fading and shadowing, both its amplitude and phase fluctuate over time causing spatio-time variation of the channel. When randomly delayed, reflected, scattered, and diffracted signal components combine constructively and/or destructively, yielding to short-term signal variations results in Multipath Fading. Slow variations of the mean signal level due to structures and components inside buildings, result in Shadowing. These spatio-time variations of the signal in a wireless environment can be characterized by the following parameters:

- Distribution of Arrival Time Sequence

- Distribution of Path Amplitudes

- Distribution of Path Phases

- Interdependence of Path Variables

2.3.2   Temporal Variation of the Channel

The propagation environment is non-stationary over time due to
the motion of the transmitter and/or the receiver as well as changes
and/or motion of the surrounding objects. With a few exceptions,
most experimental measurements in literature consider the communication channel to be stationary, or quasi-stationary or to be
wide-sense stationary. Data is accumulated over short intervals of
time. Assumption of such an environment is reasonable in residential, office, or laboratory buildings, but fails to account for
crowded arenas like shopping malls, supermarkets, or indoor sports
stadiums. In the latter case, distortions caused due to crowded
surroundings, complicated in-building structures, and continuous
motion of large number of people result in temporal variation of
the channel.

2.3.3   Large-Scale Path loss

The communication channel characteristics can change acutely,
due to increase in distance between the transmitting and receiving antenna (in SISO systems) or increase in antenna spacing (in
MIMO systems). Increase in distance makes the communication
channel vulnerable to increased number of intervening obstacles
and scatterers. Path loss is defined as the measure of the attenuation of the received signal inflicted by this kind of channel variations. Path loss in indoor environment is also imparted by absorption losses and hence depends on the construction material and
contour of a building. The communication channel even suffers
from larger variations of path losses, than outdoor mobile channel over very short distances. Modeling path loss is challenging
but is crucial in determining the radio coverage area and optimum
location of transmitting antennas.

2.3.4 Delay Spread

Due to multipath fading, multiple copies of the transmitted signal are received at the receiver at different instants of time. As different components of the signal do not arrive simultaneously, the original signal gets spread over the time domain, termed as Delay Spread. If the period of the baseband signal exceeds the delay spread, inter-symbol interference (ISI) is a direct consequence. Mean excess delay is the measure of the average delay spread caused due to multipath components, while rms (root-mean-square) delay spread is the standard deviation of the delay of reflections, weighted proportional to the energy in the reflected wave.

2.3.5 Frequency Dependence of Channel Statistics

The wireless channel characteristics can alter drastically with variation in the frequency of operation. Different measurement campaigns over the literature yield contradictory and ambiguous results. Dissimilar environment imposes different changes in channel statistics over different frequencies of operation, since path loss and continuous wave penetration loss depend on the building material used and the inner contour of the building. The delay spread or the power delay profile decay exponent in turn also exhibits different variabilities depending on the building properties. Recent investigations over potential 60GHz communication revealed that an additional attenuation of 15dB/Km is inflicted for this frequency of operation. Hence, predicting wireless propagation behavior over a wide range of frequencies is extremely complicated.

2.3.6 Noise and Co-Channel Interference

Some unwanted factors like interference, noise, inter-symbol interference (ISI), and intermodulation products also degrade the system performance of wireless radio communication. Introduction of MIMO systems and spatial reuse of radio spectrum have resulted in co-channel interference. When two transmitters separated by a certain distance transmit over identical carrier frequency, co-channel interference will result. Both the desired and the interfering signals may travel to the receiver through the same or different paths; in the latter case, the two signals fade independently with or without identical distributions. Impact of co-channel interference can be measured through the signal to interference-plus-noise ratio (SINR).

2.4   Modeling Approaches

Different modeling approaches have been used over the years to characterize radiowave propagation environment.

2.4.1   Channel Impulse Response

Impulse response of a radio propagation channel can characterize its complicated random and time-varying nature. Exciting the communication channel with an impulse or a wide-band probing signal is used to measure the channel impulse response. A carrier modulated long pseudo random noise (PN) sequence, consisting of short duration (τ_l) pulses, is transmitted over the channel. At the receiver, the demodulated signal is correlated with a locally generated PN sequence and integrated over time, to obtain the impulse response as

$$h(t) = \sum_{i=1}^{N-1} A_i \delta(t - \tau_i) \tag{2.1}$$

where N is the number of multipath components, $\delta(t)$ is the unit impulse, A_i is the amplitude of the received impulse, and τ_i is the delay. Employing Fourier Transform of $h(t)$ yields the channel transfer function, given by

$$H(f) = \sum_{i=1}^{N-1} A_i \delta e^{-j\omega \tau_i} \tag{2.2}$$

If the magnitude response is constant over the signal bandwidth and the phase response is linear, the transmission channel can be assumed to be distortionless.

2.4.1.1   Analytical Modeling

Turin's Model - The first time-invariant impulse response model was first proposed by G. L. Turin [221], incorporating multipath fading components.

$$h(t) = \sum_{i=1}^{N-1} A_i \delta(t - \tau_i) e^{j\theta_i} \tag{2.3}$$

where θ_i is the phase of the received impulse. Hence, we can also write

$$h(t) = \sum_{i=1}^{N-1} A_i \delta(t - \tau_i) e^{j2\pi f_0 (t - \tau_i)} \tag{2.4}$$

where f_0 is the carrier frequency. Now, if a low-pass signal $x(t)$, which does not vary appreciably over intervals less than τ_l, is transmitted, the received signal can be given by

$$y(t) = \int_{-\infty}^{\infty} x(\tau) h(t - \tau) \, d\tau + n(t) \tag{2.5}$$

where $n(t)$ is the complex Additive White Gaussian Noise (AWGN). Therefore

$$y(t) = Re\{[\sum_{i=1}^{N-1} A_i x(t - \tau_i) e^{j2\pi f_0 (t - \tau_i)} + n(t)].e^{j2\pi f_0 t}\} \tag{2.6}$$

Turin's model is an effective means of deducing narrowband channel models. If we put $x(t) = 1$ in (5), the resultant continuous wave envelope (R) and phase (ϕ) for a single point in space can be given by

$$Re^{j\phi} = \sum_{i=0}^{\infty} A_i e^{j2\pi f_0 (t - \tau_i)} \tag{2.7}$$

Wideband impulse response can also be used to generate the narrowband channel model by sampling the impulse response at regular internals.

2.4.1.2   Physical Modeling

Saleh's Model—In a physical channel model, the multipath components (MPCs) are modeled separately including information of their time of arrival (TOA) and departure (TOD) and their angle of arrival (AOA) and departure (AOD). Hence, the channel impulse response (CIR) models are based on the number of paths of arrival, their magnitudes, times of incidence, and locations. Saleh [85, 86, 87, 88] proposed a statistical path arrival model based on the fact that the measured rays arrive in clusters. The arrival process of the first rays of successive clusters is modeled using a Poisson process with a fixed mean arrival rate of Λ. Rest of the rays within each cluster are assumed to arrive following a Poisson process with fixed mean arrival rate of λ such that $\lambda \gg \Lambda$. Consider T_l is the arrival time of the lth cluster ($l = 0, 1, 2 \ldots$), τ_{kl} is the arrival time of the kth ray of the lth cluster ($k = 0, 1, 2 \ldots$), $T_0 = 0$, and $\tau_{0l} = 0$ for the first cluster. Then, the inter-arrival probability density functions (PDFs) of consecutive clusters and successive rays within the same cluster can be given by

$$P(T_l|T_{l-1}) = \Lambda e^{-\Lambda(T_l - T_{l-1})} \qquad l > 0 \qquad (2.8)$$

and

$$P(\tau_{kl}|\tau_{(k-1)l}) = \lambda e^{-\lambda(\tau_{kl} - \tau_{(k-1)l})} \qquad k > 0 \qquad (2.9)$$

The cluster delay times and the delay of single MPC relative to the cluster are both modeled with Poisson processes. The average cluster amplitude as well as the MPCs in each cluster is modeled by exponential decaying functions. Saleh's model has been modified several times over literature, to predict path loss, time-invariant impulse response, and the RMS delay spread and provide detailed ray tracing techniques for propagation modeling.

2.4.2   Discrete-Time Impulse Response

To simplify simulation, continuous channel impulse response can be replaced by discrete-time impulse response model. The entire time axis is divided into small time intervals ("bins"), each containing either only one or no multipath component. Choosing the bin resolution is a major factor for the implementation of the model.

Each bin should be small enough only to capture one distinct multipath. Hence the channel impulse response can be represented as a sequence of "1"s (path present) and "0"s (path absent). Each "1" will have an amplitude and a phase value. It can be used for analysis of both narrowband and wideband [79, 213] communication systems.

2.4.2.1   Analytical Modeling

Brownian Bridge Process—As wideband spread spectrum techniques are becoming popular means of wireless communication, wideband channel modeling is gaining significance. Wideband channel suffers from rich multipath reflection losses. To characterize such a communication environment, a novel wideband multipath channel model has been proposed in [222], coined as bounded Brownian bridge model (BBBM). In BBBM, the multipaths are treated as trajectories of the Brownian bridge process. Each effective signal path is assumed to be a sample function of the Brownian bridge process, $B_{0,r_0}^{T,r_1}(t,\omega)$, where r_0 and r_1 are the coordinates of the transmitter and the receiver, t is the time index, and ω is the sample in state space. A Brownian bridge process is defined as

$$B_{0,r_0}^{T,r_1}(t,\omega) = r_0 + G(t,\omega) - \frac{t}{T}(W(T,\omega) - r_1 + r_0) \quad 0 \le t \le T \quad (2.10)$$

where $G(t,\omega)$ is a Gaussian process, called the Brownian motion and satisfies, $B_{0,r_0}^{0,r_1}(0,\omega) = r_0$ and $B_{0,r_0}^{0,r_1}(T,\omega) = r_1$. The energy loss of the received signal is imparted to the propagation loss and the reflection loss. Assuming only two phases, 0 and π, the discrete-time channel impulse response is given by

$$h(t) = \sum_{n=0}^{N}\sum_{m=1}^{M_n} P_{nm} \overbrace{\left(\prod_{k=0}^{n}\gamma_{nmk}^{1/2}\left(\frac{d_{nm}}{d_{01}}\right)^{-n/2}\right)}^{\text{propagation loss}} \cdot \delta\underbrace{\left(t - \frac{d_{nm} - d_{01}}{c}\right)}_{\text{reflection loss}}$$

$$(2.11)$$

where $k = (0,1,....n)$ represents the kth reflection, N is the total number of reflections, and M_n is the total number of multipaths, each suffering n times reflections, d_{nm} is the trajectory of a multi-path component, d_{01} is the LOS distance between the transmitter and the receiver, γ_{nmk} is the reflection loss co-efficient, n is the propagation loss exponent, p_{nm} is the measure of signal inversion due to reflection, and c is the free-space velocity of light.

2.4.3 Ray Tracing

Analytical ray-tracing techniques have been proposed for radio propagation modeling over literature [39, 42, 68, 129, 166, 167, 176]. This technique has been proposed to predict path loss, the time-invariant impulse, response and the RMS delay spread. Reliable site-specific propagation prediction models for each building based on its detailed geometry and construction can be used as very effective tools in engineering communication systems. A finite number of reflectors with known location and dielectric properties are assumed for ray tracing. Maxwell's equation with boundary conditions is used to solve for the properties of the multipath components. Ray tracing yields best results when the receiver is many wavelengths away from the nearest scatterer and all the scatterers are large relative to the wavelength.

2.4.3.1 Analytical Modeling

Two Ray Model—This is the simplest form of ray tracing accounting for one LOS path or ray and only one multipath component ray, reflected off the ground (refer to Fig. 2.1). Ignoring the surface wave attenuation, the received signal, by superposition, can be given by

$$y_{2-ray}(t) = Re\left\{\frac{\lambda}{4\pi}\left[\frac{\sqrt{G_l}u(t)e^{-j2\pi l/\lambda}}{l} + \frac{R\sqrt{G_r}u(t-\tau)e^{-j2\pi(x+x')/\lambda}}{x+x'}\right]\right.$$
$$\left. \cdot e^{j2\pi f_o t}\right\} \qquad (2.12)$$

where, $\tau = (x + x' - l)/c$ is the time delay of the ground reflection relative to the LOS ray, $\sqrt{G_l} = \sqrt{G_a G_b}$ is the product of the transmit and receive antenna field radiation patterns in the LOS direction, R is the ground reflection coefficient, and $\sqrt{G_r} = \sqrt{G_c G_d}$ is the product of the transmit and receive antenna field radiation patterns corresponding to the rays of length x and x', respectively.

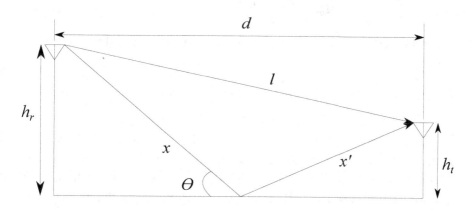

FIGURE 2.1
Two-Ray Model

2.4.4 Channel Transfer Matrix

Accommodating large number of users with higher bandwidth requirement for reliable transmission of different multimedia applications has resulted in the ever growing demand for wireless mobile networks with high capacity. In the last decade, the world witnessed the emergence and the incredible expansion in the field of MIMO systems. MIMO systems facilitate multiple users to communicate over the same time and space. It increases link capacity and throughput, improves the quality of service (QoS), and exploits diversity to improve performance to a substantial extent. Characterizing a MIMO channel can be achieved through the modeling of the channel transfer matrix (CTM) instead of individual channel impulse responses. Let us consider a MIMO system consisting of M_t transmitting antennas and M_r receiving antennas. In an environment, the communication channel can be considered as narrow-band, flat-fading, and quasi-time invariant. The communication link between each pair of transmit and receive antennas can be represented by a unique user-specific channel transfer function (CTF), $h_{p,m}$, associating the mth transmit antenna with the pth

receive antenna. Hence, the CTF for the mth user can be given by

$$\mathbf{h}_m = [h_{1,m} \ h_{2,m} \ \cdots \ h_{M_r,m}]^T \qquad (2.13)$$

The received signal at the pth antenna will be given by

$$\mathbf{y}_p(k) = \sum_{m=1}^{M_t} \mathbf{h}_{p,m} \mathbf{s}_m(k) + \mathbf{n}_p(k) \qquad (2.14)$$

where k is the symbol index, $\mathbf{s}_m(k)$ is the kth transmitted symbol from the mth transmit antenna, while $\mathbf{n}_p(k)$ is the complex-valued Additive White Gaussian Noise (AWGN) associated with the kth symbol. The overall MIMO system can then be modeled as

$$\mathbf{y}(k) = \mathbf{H}\mathbf{s}(k) + \mathbf{n}(k) \qquad (2.15)$$

where the CTF matrix, $\mathbf{H} \in \mathbb{C}^{M_r \times M_t}$ can be defined by

$$\mathbf{H} = [\mathbf{h}_{p,m}] \quad 1 \le p \le M_r \text{ and } 1 \le m \le M_t \qquad (2.16)$$

and $\mathrm{E}[|\mathbf{h}_{p,m}|^2] = 1$. If $\mathbf{s}(k) \in \mathbb{C}^{M_t \times 1}$ is the transmit symbol column vector with $\mathrm{E}[|\mathbf{s}_m(k)|^2] = \sigma_s^2$, then

$$\mathbf{s}(k) = [s_1(k) \ s_2(k) \ \cdots \ s_{M_t}(k)]^T \qquad (2.17)$$

$\mathbf{y}(k) \in \mathbb{C}^{M_r \times 1}$ denotes the receive symbol column vector

$$\mathbf{y}(k) = [y_1(k) \ y_2(k) \ \cdots \ y_{M_r}(k)]^T \qquad (2.18)$$

and $\mathbf{n}(k) \in \mathbb{C}^{M_r \times 1}$ is the complex-valued AWGN column vector with $\mathrm{E}[\mathbf{n}\mathbf{n}^H] = 2\sigma_n^2 \mathbf{I}_{M_r}$ and $\mathbf{I}_{M_r}$ being $M_r \times M_r$ identity matrix.

$$\mathbf{n}(k) = [n_1(k) \ n_2(k) \ \cdots \ n_{M_r}(k)]^T \qquad (2.19)$$

A variety of MIMO channel models have been proposed, which can be broadly classified as physical and analytical channel models. Analytical MIMO channel models are also sometimes called tensor-based models and the channel properties are fixed by the measurement scenario. A few MIMO channel models have been summarized below.

2.4.4.1   Analytical Modeling

Kronecker Model—Consider a MIMO system with transmit correlation matrix, $\mathbf{R_T}$ and receive correlation matrix $\mathbf{R_R}$. The frequency flat narrowband channel transfer matrix is given by $\mathbf{H} \in \mathbb{C}^{M_r X M_t}$, where M_t is the number of transmit and M_r is the number of receive antennas. Consider a correlation matrix $\mathbf{R_H}$, formed by the Kronecker product of the transmit and receive correlation matrices.

$$\mathbf{R_H} = \frac{1}{tr\{\mathbf{R_T}\}}\mathbf{R_T}^T \otimes \mathbf{R_R} \tag{2.20}$$

such that $\mathbf{R_T} = E_H\{\mathbf{H}^H\mathbf{H}\}$ and $\mathbf{R_R} = E_H\{\mathbf{H}\mathbf{H}^H\}$. Then $\mathbf{R_H}$ can be calculated as

$$\mathbf{R_H} = E_H\{vec\{\mathbf{H}\}.vec\{\mathbf{H}\}^H\} \tag{2.21}$$

where $vec\{\cdot\}$ converts a matrix into a column vector. [225] The MIMO channel transfer matrix can then be modeled as

$$\mathbf{H} = \frac{1}{tr\{\mathbf{R_T}\}}\mathbf{R_R}^{1/2}\mathbf{G}(\mathbf{R_T}^{1/2})^T \tag{2.22}$$

where $\mathbf{G}$ is an iid complex Gaussian distributed symmetric fading matrix, with each entry having zero mean and unit variance. For any MIMO system, decomposition of the correlation matrix must be possible according to (1). This is applied as a constraint for employing the Kronecker Model, which sometimes makes it impossible to implement for the direct LOS propagation.

Rician Model—The Rician Channel Model [226] divides the channel transfer matrix into two parts: one is the correlated LOS path $\mathbf{H_{LOS}}$ and other is the NLOS path $\mathbf{H_{NLOS}}$. The channel transfer matrix can be expressed as

$$\mathbf{H} = \sqrt{\frac{\kappa}{1+\kappa}}\mathbf{H_{LOS}} + \sqrt{\frac{1}{1+\kappa}}\mathbf{H_{NLOS}} \tag{2.23}$$

where κ is the Rician K-factor. The Rician K-factor is defined as the ratio of the signal power over the direct LOS propagation path to the signal power over the indirect NLOS path.

Weichselberger Model—Weichselberger Model was proposed in [227] as a modification of the Kronecker Model. A new model parameter called coupling matrix $\boldsymbol{\Omega}$ was introduced to make the correlation matrices $\mathbf{R_T}$ and $\mathbf{R_R}$ easily decomposable in terms of their eigenbasis and eigenvalues

$$\mathbf{R_T} = \mathbf{U_T}\boldsymbol{\Lambda_T}\mathbf{U_T}^{H} \quad \text{and} \quad \mathbf{R_R} = \mathbf{U_R}\boldsymbol{\Lambda_R}\mathbf{U_R}^{H} \tag{2.24}$$

where $\boldsymbol{\Lambda_T}$ and $\boldsymbol{\Lambda_R}$ are eigenvalue matrices and $\mathbf{U_T}$ and $\mathbf{U_R}$ are eigenbasis matrices of $\mathbf{R_T}$ and $\mathbf{R_R}$, respectively. The channel transfer matrix, in turn will be given by

$$\mathbf{H} = \mathbf{U_R}(\tilde{\boldsymbol{\Omega}} \odot G)\mathbf{U_T}^{T} \tag{2.25}$$

where $\tilde{\boldsymbol{\Omega}}$ denotes the element wise square root of $\boldsymbol{\Omega}$ and $\boldsymbol{\Omega}$ is calculated from

$$\boldsymbol{\Omega} = E_H\{(\mathbf{U_R}^{H}\mathbf{H}\mathbf{U_T}^{*}) \odot (\mathbf{U_R}^{T}\mathbf{H}^{*}\mathbf{U_T})\} \tag{2.26}$$

2.4.4.2   Physical Modeling

Extended Saleh Model (ESM) - Saleh's model [228] proposed cluster-based MPC stochastic modeling for SISO channels. It has been extended to a MIMO channel model by the authors in [229]. Laplacian probability distribution has been used to model the AOAs and AODs of the MPCs. A single CIR has been extended to $M_t \times M_r$ CIRs using the path length differences of all MPCs originating between neighboring antennas. Antenna set-up, AOA, and AOD have been used as the primary modeling parameters, assuming planar wave fronts.

Extended Saleh Spherical Wave Model (ESSWM)—Another modification of the Saleh Model was proposed in [224]. The LOS path is directly calculated from the presumed antenna set-up. The AOA and AOD for the MPCs are modeled by uniform distribution based on the accumulated measurement. Hence, a-priori knowledge of the antenna arrangement and angular profiles at the transmitter and receiver are necessary for application of this channel model. The SISO model is extended to its MIMO version with $M_t \times M_r$ CIRs calculated from path length differences of all MPCs occurring between neighboring antennas.

Zwick's Channel Model—A novel channel model for wireless environment was introduced by Zwick et al. in [230]. Appearance and disappearance of MPCs are modeled by a marked Poisson process. The transitions between LOS and obstructed LOS components are modeled by Markov process. The AOAs and AODs are modeled by Laplacian distribution and uniform distribution over large delay times. The SISO model can be extended to a MIMO CTM based modeling assuming planar wave fronts.

2.5 Modeling Spatio-Temporal Variations

EM waves consist of time-varying electric and magnetic fields propagating in the media. In their simple form, radiowaves are time harmonic fields in sinusoidal form with a frequency. When the medium is not dispersive, its velocity is not dependent on the frequency but is only related to the medium parameters.

2.5.1 Distribution of Arrival Time Sequence

The wireless communication scenario can be assumed as a concoction of few LOS propagation paths and several NLOS propagation paths. Each signal can arrive through these paths, at single or multiple receivers at different instants of time. the entire sequence of multiple arrival times $\{t_k\}_0^\infty$ can be assumed to form a point process on the positive time axis. The LOS paths however suffer a fixed delay t_0 and hence are excluded from the sequence. The resultant arrival time sequence can thus be modeled by characterizing $\{t_k - t_0\}_1^\infty$. In the following subsections we will be summarizing some of the distributions proposed over years, to model the arrival process and arrival time sequence.

2.5.1.1   Standard Poisson Model

The standard Poisson model assumes that MPCs are generated as a result of randomly located obstacles and intervening objects. Hence, within a time duration of T, the number of paths l originate with a probability

$$P(l) = \frac{\mu^l e^{-\mu l}}{l!} \tag{2.27}$$

where μ is the Poisson parameter and is related to the mean arrival rate, $\lambda(t)$ at time t, by the relation

$$\mu = \int_T \lambda(t)\,\mathrm{d}t \tag{2.28}$$

For a standard and/or stationary Poisson process, interarrival times between successive MPCs are independent identically distributed random variables with exponential decaying distribution,

$$f_X(x) = \lambda(t)e^{-\lambda x} \qquad x > 0 \tag{2.29}$$

where $x_i = t_i - t_{i-1}$ is the interarrival time for $i = 0, 1, 2 \ldots$.

But standard Poisson distribution has failed to provide a good fit for measurement data collected in different environments in [56, 57, 78, 79, 85, 86, 148]. The deviations can be attributed to the fact that the obstacles inside a building do not always impart totally random dispersion and scattering patterns. Hence, several other distributions for arrival time sequences have been proposed to provide a better fit to collected measurements.

2.5.1.2   Modified Poisson—The $\Delta - K$ Model

This model was proposed as a modification over the standard Poisson process by Turin et al. in [221], but was developed to full potential by Suzuki in [213]. This model takes into account the possibility of MPCs arriving in clusters. The communication channel is represented by two states, S_1 and S_2, with mean arrival rates λ and $K\lambda$, respectively. Initially, the process is assumed to start in state S_1. If a path arrives at time t, the state of the process changes to S_2, for a time period of $(t, t + \Delta)$. If no path arrives during this interval, the process switches back to state S_1. For $K = 1$, or

$\Delta = 0$, the $\Delta - K$ model reduces to a standard Poisson process. For $K > 1$, the process exhibits clustering, while for $K < 1$, the paths tend to arrive at evenly spaced intervals.

The $\Delta - K$ model has been shown to exhibit good fit to measurement data collected in several buildings [56, 57], dissimilar office buildings [78, 79], and factory environments [33]. However, success of the model is dependent on the choice of K as well as individual small intervals of length Δ.

2.5.1.3   Modified Poisson—Non-exponential interarrivals

A statistical model for CIR was proposed in [148, 154] based on detailed accumulated data in several factory environments [23, 26, 27]. Weibull distribution was applied as a good fit to the interarrival time measurements between MPCs. Three parameters, the shape parameter, the scale parameter, and the location parameter, are used to offer flexibility in modeling. If only one of the parameter is varied, Weibull distribution reduces to exponential distribution and the arrival time sequence can be modeled by a standard Poisson process.

2.5.1.4   Neymann-Scott Clustering Model

This model is based on double Poisson process and is used in cosmology to study distribution of galaxies. Measurement collected in office buildings [85, 86] has been used to fit this model. Clustering of multipath components is assumed to be caused by stationary large building structure components like walls and doors. Temporary variations of immediate environment contribute to the multiple reflected components within each cluster.

2.5.1.5   Gilbert's Burst Noise Model

Gilbert's burst noise model [217, 218] can also be used to characterize arrival time sequences in a wireless environment. The communication process can be assumed to have two states. State S_1 will account for any error-free transmission, while state S_2 will take into account errors with a pre-assigned probability. The process will switch between environments depending on the communication environment, but independent of the events.

2.5.1.6 Pseudo-Markov Model

Pseudo-Markov model was introduced in [219, 220] to identify spike trains from nerve cells. The process can switch between two states S_1 and S_2 with different interarrival distributions x_1 and x_2. If N_1 and N_2 are random variables with a given distribution, after N_1 events occurring in state S_1, the process switches to state S_2. Similarly after N_2 events have occurred the process switches back to state S_1 from S_2.

2.5.1.7 α-Stable Distribution Model

A statistical model for the path arrival time in 60-GHz wideband communication channel is proposed in [223]. This modeling is based on the class of α-stable distributions. Though there is no exact characteristic function can be approximated as

$$\phi(\theta) = \begin{cases} exp\{-\sigma^\alpha|\theta|^\alpha(1 - i\beta sign(\theta)tan\frac{\pi\alpha}{2}) + i\mu\theta\} & \text{if } \alpha \neq 1 \\ exp\{-\sigma|\theta|(1 + i\beta\frac{2}{\pi}sign(\theta)ln|\theta|) + i\mu\theta\} & \text{if } \alpha = 1 \end{cases}$$

(2.30)

where

$$sign(\theta) = \begin{cases} 1 & \text{if } \theta > 0 \\ 0 & \text{if } \theta = 0 \\ -1 & \text{if } \theta < 0 \end{cases}$$

(2.31)

α is the characteristic exponent $(0 < \alpha \leq 2)$ measuring the thickness of the distribution tail, μ is the location parameter $(\infty < \mu < \infty)$, σ is the dispersion parameter $(\sigma > 0)$ and β is the index of symmetry $(-1 \leq \beta \leq 1)$ about its central location. Now, consider N be the number of obstacles between the transmitter and the receiver, $d_{k,i}$ be the delay between reflection on the kth and the $(k + 1)^{th}$ reflectors for a given path i and T be the total transmission time. If the first ray arrives at time t_0, the ith ray will arrive with a delay

$$\tau_i = \sum_{k=1}^{N_i} d_{k,i} - t_0$$

(2.32)

and the interarrival delay $d_{k,i}$ between two successive obstacles can be distributed as

$$P(d_{k,i} > x) = \begin{cases} \delta_i^\alpha x^{-\alpha} & \text{if } x > 1/\delta_i \\ 1 & \text{if } x \leq 1/\delta_i \end{cases} \qquad (2.33)$$

where δ_i^α is the average number of obstacles encountered by the ith ray over a duration of δ_i. If $0 < \alpha < 1$, and δ_i increases to infinity, τ_i converges in an α-stable distribution with a scale parameter, $\sigma^\alpha = \Gamma(1 - \alpha)Cos(\frac{\pi\alpha}{2}) > 0$. If $1 < \alpha < 2$, and δ_i increases to infinity, $\tau_i - \frac{\alpha}{\alpha-1}[\frac{1}{\delta_i}]^{1-\alpha}$ converges in an α-stable distribution with a scale parameter, $\sigma^\alpha = \Gamma(1 - \alpha)Cos(\frac{\pi\alpha}{2}) > 0$, where Γ is the gamma function.

These distributions appear to offer a good fit for arrival time sequence in 60-GHz communication channel measurements. However, applicability of the model to different environments and other frequency bands has to be tested before widespread acceptance.

2.5.2 Distribution of Path Amplitudes

If the transmitter or the receiver is in motion or the scattering environment in space changes, the spatial variations in amplitude and phase of the received signal will manifest themselves as time variations. As radio channels are reciprocal in nature, the propagation link carries equal energy in both directions. However, the spatial distribution of the arriving rays will be different in each direction. In a wireless environment, presence of scatterers may result in rays arriving from different directions without any direct LOS path. It can also give rise to a dominant LOS path or a dominant reflected path accompanied by several secondary weaker NLOS propagation paths. Large in-building structures like doors and walls can even result in slow variations of the propagating signal amplitudes. Depending on the communication environment, these amplitude variations can be extended to narrowband or wideband temporal fading. The former is used to model scenarios when both transmit and receive antennas are fixed, while the later is used to model variations caused by the motion of surrounding scatterers, like people or equipment inside the building. Hence, several distributions have been proposed to capture the effect of amplitude variations over different propagation scenarios, a few of which will be summarized in this subsection.

2.5.2.1 Rayleigh Distribution

Multipath fading is most widely modeled using Rayleigh distribution, when there is no direct LOS path between the transmitter and the receiver. The fading amplitude, α is distributed according to

$$p_\alpha(\alpha) = \frac{2\alpha}{\Omega} \, exp\left(-\frac{\alpha^2}{\Omega}\right) \qquad \alpha \geq 0 \qquad (2.34)$$

where Ω is defined as the mean-square value of the random variable. The instantaneous signal to noise ratio (SNR) per symbol of the channel γ is distributed according to exponential distribution given by

$$p_\gamma(\gamma) = \frac{1}{\bar{\gamma}} \, exp\left(-\frac{\gamma}{\bar{\gamma}}\right) \qquad \gamma \geq 0 \qquad (2.35)$$

where $\bar{\gamma} = E\{\gamma\}$ denotes the average SNR. The Moment Genarating Function (MGF) corresponding to this fading model is given by

$$\mathcal{M}_\gamma(s) = (1 - s\bar{\gamma})^{-1} \qquad (2.36)$$

and the moments associated with this fading model can be expressed by

$$E(\gamma^k) = \Gamma(1 + k)\bar{\gamma}^k \qquad (2.37)$$

The Autocorrelation Function (AF) for a Rayleigh fading model is equal to 1 and therefore typically offers a good fit for experimental data in wireless environments, where there is no direct LOS path. Measurements have been taken in factory environments [94, 95], office buildings [85, 86], university campus building [123], both in wideband [17, 18] and narrowband continuous wave [2, 3, 18] transmission from outside the building or within the building, where Rayleigh distribution has been efficiently applied to model the amplitude variations.

2.5.2.2 Rician Distribution

Rician distribution was introduced in [216] to model propagation paths consisting of one strong direct LOS component and several random weaker NLOS components. The channel fading amplitude follows the following distribution.

$$p_\alpha(\alpha) = \frac{2(1+n^2)e^{-n^2}\alpha}{\Omega} \, exp \left(-\frac{(1+n^2)\alpha^2}{\Omega} \right) \cdot I_0 \left(2n\alpha\sqrt{\frac{1+n^2}{\Omega}} \right)$$

$$\alpha \geq 0$$

$$(2.38)$$

where $I_0(\cdot)$ is the zeroth-order-modified Bessel function of the first kind and n is the fading parameter. The fading parameter is related to the Rician K-factor, given by $K = n^2$. The K-factor corresponds to the ratio of the power of the LOS component to the average power of the NLOS components. The instantaneous SNR per symbol of the channel is distributed according to

$$p_\gamma(\gamma) = \frac{(1+n^2)e^{-n^2}}{\bar{\gamma}} \, exp \left(-\frac{(1+n^2)\gamma}{\bar{\gamma}} \right) \cdot I_0 \left(2n\sqrt{\frac{(1+n^2)\gamma}{\bar{\gamma}}} \right)$$

$$\gamma \geq 0$$

$$(2.39)$$

The MGF is given by

$$\mathcal{M}_\gamma(s) = \frac{1+n^2}{(1+n^2) - s\bar{\gamma}} \, exp \left(\frac{n^2 s\bar{\gamma}}{(1+n^2) - s\bar{\gamma}} \right) \qquad (2.40)$$

with moments calculated from

$$E(\gamma^k) = \frac{\Gamma(1+k)}{(1+n^2)^k} \, {}_1F_1(-k, 1; -n^2)\bar{\gamma}^k \qquad (2.41)$$

where ${}_1F_1(\cdot, \cdot; \cdot)$ is the Kummer confluent hyper-geometric function. The Rician distribution covers the range from worst case Rayleigh fading $(n = 0)$ to best case no fading $(n = \infty)$. Rician distribution exhibits good agreement with measured data [69, 70, 71] in factory environments [24, 27] and several in-building scenarios using both leaky feeders and dipole antennas [149, 151].

2.5.2.3 Hoyt Distribution

Amplitude variation modeling spanning from Gaussian distribution to Rayleigh distribution was proposed by Hoyt. Fading amplitude over Hoyt distribution is given by

$$p_\alpha(\alpha) = \frac{(1+q^2)\alpha}{q\Omega} \; exp\left(-\frac{(1+q^2)\alpha^2}{4q^2\Omega}\right) \cdot I_0\left(\frac{(1-q^4)\alpha^2}{4q^2\Omega}\right) \quad \alpha \geq 0 \tag{2.42}$$

where q is the fading parameter and $0 \leq q \leq 1$. The instantaneous SNR per symbol of the channel is distributed according to

$$p_\gamma(\gamma) = \frac{(1+q^2)}{2q\bar\gamma} \; exp\left(-\frac{(1+q^2)\gamma}{4q^2\bar\gamma}\right) \cdot I_0\left(\frac{(1-q^4)\gamma}{4q^2\bar\gamma}\right) \quad \gamma \geq 0 \tag{2.43}$$

The MGF is given by

$$\mathcal{M}_\gamma(s) = \left(1 - 2s\bar\gamma + \frac{(2s\bar\gamma)^2 q^2}{(1+q^2)^2}\right)^{-1/2} \tag{2.44}$$

with moments associated is given by

$$E(\gamma^k) = \Gamma(1+k) \; {}_2F_1\left(-\frac{k-1}{2}, -\frac{k}{2}; 1; -\left(\frac{1-q^2}{1+q^2}\right)^2\right)\bar\gamma^k \tag{2.45}$$

where ${}_2F_1(\cdot,\cdot;\cdot;\cdot)$ is the Gauss hyper-geometric function. The Hoyt distribution spans from worst case one-sided Gaussian fading ($q = 0$) to best case Rayleigh fading ($q = 1$).

2.5.2.4 Nakagami-m Distribution

Nakagami-m distribution is an essentially central chi-squared distribution and the fading amplitude is distributed as

$$p_\alpha(\alpha) = \frac{2m^m \alpha^{2m-1}}{\Omega^m \Gamma(m)} \; exp\left(-\frac{m\alpha^2}{\Omega}\right) \quad \alpha \geq 0 \tag{2.46}$$

where m is the fading parameter and $1/2 \leq m \leq \infty$. The SNR per symbol is distributed as

$$p_\gamma(\gamma) = \frac{m^m \gamma^{m-1}}{\bar\gamma^m \Gamma(m)} \; exp\left(-\frac{m\gamma}{\bar\gamma}\right) \quad \gamma \geq 0 \tag{2.47}$$

The MGF will be given by

$$M_\gamma(s) = \left(1 - \frac{s\bar{\gamma}}{m}\right)^{-m} \qquad (2.48)$$

with moments calculated from

$$E(\gamma^k) = \frac{\Gamma(m+k)}{\Gamma(m)\,m^k}\,\bar{\gamma}^k \qquad (2.49)$$

Nakagami-m distribution can account for several other fading distributions as special cases. The fading parameter m can be mapped to q-parameter of Hoyt distribution by

$$m = \frac{(1+q^2)^2}{2(1+2q^4)} \qquad m \le 1 \qquad (2.50)$$

and to n-parameter of Rician distribution by

$$m = \frac{(1+n^2)^2}{1+2n^2} \qquad n \ge 0 \qquad (2.51)$$

Simulations of continuous wave envelope variations based on analytical ray tracing techniques have exhibited that fast fading components follow Nakagami-m distribution [176].

2.5.2.5 Weibull Distribution

The fading amplitude is distributed according to

$$p_\alpha(\alpha) = c\left(\frac{\Gamma(1+2/c)}{\Omega}\right)^{c/2} \alpha^{c-1} exp\left[-\left(\frac{\alpha^2}{\Omega}\Gamma\left(1+\frac{2}{c}\right)\right)^{c/2}\right] \qquad (2.52)$$

where c is the shape parameter offering flexibility to a good fit to measurements. The instantaneous SNR per symbol will be distributed as

$$p_\gamma(\gamma) = \frac{c}{2}\left(\frac{\Gamma(1+2/c)}{\bar{\gamma}}\right)^{c/2} \gamma^{c/2-1} exp\left[-\left(\frac{\gamma}{\bar{\gamma}}\Gamma\left(1+\frac{2}{c}\right)\right)^{c/2}\right] \qquad (2.53)$$

The MGF will be given by

$$M_\alpha(s) = c\left(\frac{\Gamma(1+2/c)}{\Omega}\right)^{c/2} (2\pi)^{\frac{1-c}{2}} \frac{1}{\sqrt{c}}\left(-\frac{s}{c}\right)^{-c} \qquad (2.54)$$

$$\times G_{1,c}^{c,1}\left(\left(\frac{\Gamma(1+2/c)}{\Omega}\right)^{-c/2}\left(-\frac{s}{c}\right)^c \bigg|_{1,1+1/c,\dots 1+(c-1)/c}\right) \qquad (2.55)$$

where $G_{1,c}^{c,1}(\cdot)$ is the Meijer's G-function. The Weibull distribution has been applied in very few occasions over literature and has been seen to offer good fit to mobile radio [215] and $910MHz$ measurements [64].

2.5.2.6　Lognormal Distribution

In Lognormal distribution, the path SNR per symbol is given by a standard distribution of

$$p_\gamma(\gamma) = \frac{\xi}{\sqrt{2\pi}\sigma\gamma}\,exp\left[-\frac{(10log_{10}\gamma - \mu)^2}{2\sigma^2}\right] \quad (2.56)$$

where $\xi = 10/ln10 = 4.3429$ and μ(dB) and σ(dB) are the mean and the standard deviation of $10log_{10}\gamma$, respectively. The MGF is given by

$$\mathcal{M}_\gamma(s) = \frac{1}{\sqrt{\pi}}\sum_{n=1}^{N_p} H_{x_n}\,exp\left(10^{(\sqrt{2}\sigma x_n + \mu)/10}s\right) \quad (2.57)$$

where x_n are the zeros of the N_p-order Hermite polynomial and H_{x_n} are the weight factors of the N_p-order Hermite polynomial. Lognormal distribution has found good fit over several measurement scenarios like factory environments [160], transmitter outside the building [1, 4, 5], obstructed paths inside factory [23], and limited wideband measurement at several college buildings [57].

2.5.2.7　Suzuki Distribution

A mixture of Rayleigh and lognormal distributions was first proposed by Suzuki in [213]. The fading amplitude is distributed as

$$p_\alpha(\alpha) = \int_0^\infty \frac{2\alpha}{\Omega}\,exp\left(-\frac{\alpha^2}{\Omega}\right)\frac{\xi}{\sqrt{2\pi}\sigma\gamma}\cdot exp\left[-\frac{(10log_{10}\gamma - \mu)^2}{2\sigma^2}\right]d\gamma \quad (2.58)$$

Though complicated, Suzuki distribution accounts for one or more relatively strong signals arriving at the receiver as well as several weaker subpaths arriving due to multiple reflections or refractions and/or scattering. A successful implementation has been done in [214] for modeling wireless propagation.

2.5.2.8 Beckmann Distribution

The Beckmann distribution [212] is a four-parameter distribution corresponding to the envelope of two independent Gaussian random variables, each with their own mean and variance and hence includes the Rayleigh, Rician, Hoyt, and one-sided Gaussian distribution as special cases. Let X and Y be Gaussian random variables with parameters (μ_x, σ_x) and (μ_y, σ_y), respectively. The fading amplitude, $\alpha = \sqrt{X^2 + Y^2}$ will be distributed as

$$p_\alpha(\alpha) = \frac{\alpha}{2\pi\sigma_x\sigma_y} \int_0^{2\pi} exp\left[-\frac{(\alpha Cos\theta - \mu_x)^2}{2\sigma_x^2} - \frac{(\alpha Sin\theta - \mu_y)^2}{2\sigma_y^2} \right] d\theta$$

(2.59)

Another form of Beckmann distribution also exists in terms of doubly infinite series of products of modified Bessel functions of the first kind.

2.5.2.9 K-Distribution

A mixture of Rayleigh and gamma distributions was proposed in [211], coined as K-distribution. The fading amplitude follows a distribution given by

$$p_\alpha(\alpha) = \int_0^\infty \frac{2\alpha}{w} exp\left(-\frac{\alpha^2}{w}\right) \frac{\nu^\nu w^{\nu-1}}{\bar{w}^\nu \Gamma(\nu)} exp\left(-\frac{\nu w}{\bar{w}}\right) dw \quad (2.60)$$

and SNR per symbol is given by

$$p_\gamma(\gamma) = \int_0^\infty \frac{1}{w} exp\left(-\frac{\gamma}{w}\right) \frac{\nu^\nu w^{\nu-1}}{\bar{w}^\eta \Gamma(\nu)} exp\left(-\frac{\nu w}{\bar{w}}\right) dw \quad (2.61)$$

where ν and w are positive parameters accounting for the effective number of scatterers. As $\nu \to \infty$, K-distribution reduces (57) to Rayleigh distribution and (58) to gamma distribution. K-distribution is useful for modeling diverse scattering phenomena in wireless environments.

2.5.2.10 $K - \mu$ - Distribution

The fading model for the $K - \mu$ distribution [210] considers a signal composed of clusters of multipath waves propagating in a non-homogeneous environment. Within single cluster, the phases of the scattered waves are random and have similar delay times with delay-time spreads of different clusters being relatively large. It is assumed that the clusters of multipath waves have scattered waves with identical powers, and that each cluster has a dominant component with arbitrary power. Given the physical model for the $K - \mu$ distribution, the fading envelope α can be written in terms of the in-phase and quadrature components of the fading signal as

$$\alpha^2 = \sum_{i=1}^{n}(X_i + p_i)^2 + \sum_{i=1}^{n}(Y_i + q_i)^2 \qquad (2.62)$$

where X_i and Y_i are mutually independent Gaussian processes with $\bar{X}_i = \bar{Y}_i = 0$ and $\bar{X}_i^2 = \bar{Y}_i^2 = \sigma^2$, p_i and q_i are, respectively, the mean values of the in-phase and quadrature components of the multipath waves of cluster i and n is the number of clusters of multipath components. Hence, the total power of the ith cluster will be given by $\gamma_i = \alpha_i^2 = (X_i + p_i)^2 + (Y_i + q_i)^2$. The PDF for the instantaneous SNR per symbol is given by

$$p_\gamma(\gamma) = \frac{\mu(1+\kappa)^{\frac{\mu+1}{2}}\gamma^{\frac{\mu-1}{2}}}{\kappa^{\frac{\mu-1}{2}}exp(\mu\kappa)\bar{\gamma}^{\frac{\mu+1}{2}}}exp\left(-\frac{\mu(1+\kappa)\gamma}{\bar{\gamma}}\right)I_{\mu-1}\left(2\mu\sqrt{\frac{\kappa(1+\kappa)\gamma}{\bar{\gamma}}}\right) \qquad (2.63)$$

where $\mu = \frac{E^2\{\gamma\}}{var\{\gamma\}} \cdot \frac{1+2\kappa}{(1+\kappa)^2}$ and $\kappa > 0$ is the ratio of the total power of the dominant components to that of the scattered waves.

2.5.2.11 $\eta - \mu$ - Distribution

The $\eta - \mu$ distribution [209] is an arbitrary distribution, fully characterized in terms of measureable physical parameters. Therefore it is possible to fit experimental data by adequately setting the two shape parameters, η and μ. The $\eta - \mu$ distribution accounts for both Hoyt ($\mu = 0.5$) and Nakagami-m ($\eta \to 0, \eta \to \infty, \eta \to \pm1$) distributions. The PDF for the instantaneous SNR per symbol is given by

$$p_\gamma(\gamma) = \frac{2\sqrt{\pi}\mu^{\mu+1/2}h^\mu\gamma^{\mu-1/2}}{\Gamma(\mu)H^{\mu-1/2}\bar{\gamma}^{\mu+1/2}}exp\left(-\frac{2\mu\gamma h}{\bar{\gamma}}\right)I_{\mu-1/2}\left(\frac{2\mu H\gamma}{\bar{\gamma}}\right) \qquad (2.64)$$

where $\mu = \frac{E^2\{\gamma\}}{2var\{\gamma\}}\left[1 + \left(\frac{H}{h}\right)^2\right]$ and $I_x(\cdot)$ is the modified Bessel function of the first kind of order x. The parameters H and h can be specified in two different formats. In Format 1,

$$H = (\eta^{-1} - \eta)/4 \quad \text{and} \quad h = (2 + \eta^{-1} + \eta)/4 \qquad (2.65)$$

where $0 < \eta < \infty$ is the ratio of the power in in-phase component to that in quadrature scattered waves in each multipath cluster. In Format 2,

$$H = \eta/(1 - \eta^2) \quad \text{and} \quad h = 1/(1 - \eta^2) \qquad (2.66)$$

where $-1 < \eta < 1$ is the correlation coefficient between the in-phase and quadrature scattered waves in each multipath cluster.

2.5.2.12 $\alpha - \mu$ - Distribution

The $\alpha - \mu$ distribution [208] is a generalized fading distribution that can be used to represent various fading scenarios and other standard fading models as special cases. The distribution deals with non-linearity of the propagation medium. Fading signal with amplitude r can be expressed in terms of an arbitrary constant parameter $\alpha > 0$ as, $\hat{r} = \sqrt[\alpha]{E(r^\alpha)}$. The fading amplitude will have a PDF $p(r)$,

$$p(r) = \frac{\alpha\mu^\mu r^{\alpha\mu-1}}{\hat{r}^{\alpha\mu}\Gamma(\mu)} \, exp\left(-\mu\frac{r^\alpha}{\hat{r}^\alpha}\right) \qquad (2.67)$$

where μ is a shape parameter accounting for the number of multipath clusters. The $\alpha-\mu$ distribution can be used to derive Rayleigh distribution ($\alpha = 2, \mu = 1$), Nakagami-m distribution (μ is the fading parameter and $\alpha = 2$), and Weibull distribution ($\alpha/2$ is the fading parameter and $\mu = 1$).

2.5.3 Distribution of Path Phases

Envelope variation of propagating wave in wireless environment is partly imposed by the variation of statistical properties of the phase sequence, $\{\theta_k\}_0^\infty$. Though phase variation has been recognized as a very important factor in characterizing the communication channel, there is no empirical model proposed to date, to identify its distribution. This may be due to the fact that measuring phase of individual multipath components is an extremely challenging task.

As the path length changes by a single wavelength of operation, the signal phase alters by an order of 2π. Hence, a small change in the location of the transmitter or receiver or scatterers can result in an immense deviation of phase. Hence, modeling phase θ_k with an uniform distribution $U[0, 2\pi]$ is the prevalant technique in literature. However, aberrations from this uniform distribution are often evident over small sampling distances, while if sampled at symbol rate, phase values appear to be highly correlated over a fixed delay. Variable arrival time sequence of multipath components will again result in large variations in phase and successive multipath components will have independent, random, and uncorrelated phases. Hence, it is crucial to model phase changes of a propagating signal rather than the absolute phase value of a multipath component at a fixed point in space.

Let the phase of a multipath component with a fixed delay is $\theta^{(m)}$, where $m = 1, 2, 3, \ldots$ is the number of adjacent points in space in a given region. For $m = 1$, $\theta^{(m)}$ is assumed to pick values from a distribution of $U[0, 2\pi)$ and

$$\theta^{(m)} = \theta^{(m-1)} + \phi(s_m/\lambda) \qquad m = 2, 3, \ldots \qquad (2.68)$$

where s_m is the spatial separation between the $(m-1)^{th}$ and the mth set of points, λ is the operating wavelength, and $\phi(s_m/\lambda)$ will therefore represent the phase increment. This set of phase values can again reappear if and only if another multipath component arrives with the same fixed delay. Correlation of phase values can be implemented through the appropriate choice of the phase increment parameter. Modeling phase changes using this approach have been considered several times in literature. They can be broadly classified into two types, a brief summary of which is provided in the following subsections.

2.5.3.1 Random Phase Increment Model

The phase increment parameter, $\phi(s_m/\lambda)$ is considered as a random variable, which can be, for example, Gaussian distributed with zero mean and standard deviation of σ_s/λ. The initial phase is chosen from a uniform distribution $U[0, 2\pi)$ and subsequent values are obtained by adding the random increment of $\phi(s_m/\lambda)$. The degree of correlation between $\theta^{(m-1)}$ and $\theta^{(m)}$ is controlled by making σ_s/λ an increasing function of s/λ, or s for a fixed value of λ. As σ_s/λ increases, the correlation between $\theta^{(m-1)}$ and $\theta^{(m)}$ decreases until they become completely uncorrelated. Hence for $s = 0$, $\sigma_s/\lambda = 0$, and $\theta^{(m-1)} = \theta^{(m)}$. There is no guiding rule for choosing the probability distribution of $\phi(s_m/\lambda)$ and hence can be based on measurement and simulation results. However, Gaussian distribution for $\phi(s_m/\lambda)$ and exponential distribution for σ_s/λ have been used in [206, 207] for simulation of phase in wideband radio channel model.

2.5.3.2 Deterministic Phase Increment Model

The changes in the phase value of a multipath component are calculated from a known $\theta^{(1)}$ and $\phi(s_m/\lambda)$. It is assumed that within a length of one meter in space, all multipath components with the same delay are caused by reflection from the same fixed scatterer [29, 33]. The phase increment is calculated using a single scatterer and the local geometry. This model has been applied to simulate radio channel in [103] and mobile channel in [205]. In both cases, the angle of arrival of the lth multipath component is assumed to remain same with respect to the direction of motion, ψ_l of the transmitter or receiver for small separations in space, and the phase increment is given by

$$\phi(s_m, \lambda) = \frac{2\pi s_m}{\lambda} \, Cos\psi_l \qquad (2.69)$$

where ψ_l was selected from a uniform distribution. For wireless environment, ψ_l is estimated with a 5° resolution based on measurements accumulated in [102] and then applying Fourier transform method [103]. However, both the above-mentioned models do not offer realistic modeling phase in radio channel. Both random

and deterministic approaches are used to simplify phase characterization of multipath components, while the real scenario is much
more complicated as well as difficult to measure.

2.5.4 Interdependence within path variables

The correlation between the path variables can be characterized
using different techniques for different scenarios.

2.5.4.1 Correlation within a profile

Correlation is likely to exist between adjacent multipath components of the same impulse response profile, but has not yet been
explored in details in prevalent literature. Due to clustering property of local scatterers in a wireless environment, correlation exists between arrival time sequences of successive multipath components. High-resolution measurements exhibit this phenomenon
as the scattering objects remain same over a certain interval of
time. Direct LOS paths become uncorrelated with their multipath
components if the temporal separation exceeds 8 *ns* [25, 33]. For
NLOS paths to be uncorrelated, the temporal separation has to
be more than 25 *ns*. Conclusively multipath components are found
to be uncorrelated for a spatial separation larger than 3λ, where
λ is the operating wavelength. Similar results are observed for a
temporal separation larger than 100 *ns*.

2.5.4.2 Correlation between spatially separated profiles

Spatial correlation between closely spaced impulse response profiles sometimes results in large variations in amplitudes, arrival
times, and phases. For the LOS path and its reflected components
to be uncorrelated in amplitude, the temporal separation has to
exceed 100 *ns* and the spatial separation has to be more than $\lambda/2$
[27], while the NLOS path and its successive multipath components are uncorrelated or minimally correlated for all spatial and
temporal correlations in a wireless environment. Correlation of log-
amplitude of impulse response profiles of spatially close points has
been extensively studied [77, 78, 79]. For a displacement of 2 *cm* of
the mobile terminal, the correlation coefficient varies between 0.7

and 0.9. However, it decreases rapidly with higher displacements of the mobile terminal or the scatterers.

2.5.4.3 Correlation between spatial and temporal variations

In a dynamic environment, the amplitude variations may be correlated with the arrival time sequence. As subsequent multipath components suffer from larger attenuations due to larger propagation paths, there is possibility of multiple reflections. Hence correlation between the appearance and disappearance of multiple signal paths can arise and has been modeled using M-step 4-state Markov Channel Model (MCM) in [184]. Correlated Birth-Death (B-D) processes have been used to represent such a scenario. Due to the distinction in the B-D statistics, spatio-temporal dispersion, and correlation properties of different propagation scenarios, the model is generalized by segments of measurement runs with the same communication mode. However, correlation between the phase sequence and arrival time sequence and/or amplitude variations has not been examined and modeled.

2.6 Modeling Large-Scale Path loss

Several path loss measurement and modeling techniques have been proposed in literature. In a wireless environment, path loss varies rapidly over very short distances and hence is extremely difficult to model, due to its sensitivity to several factors. Some of the distinct path loss models are reviewed in the following subsections.

2.6.1 Model 1

The received signal follows an inverse exponent law with the spacing between transmit and receive antennas

$$P(d) = P_0 d^{-n} \tag{2.70}$$

where d is the distance between the transmitter and the receiver, $P(d)$ is the corresponding received power, and P_0 is the received power corresponding to $d = 1m$. Hence, P_0 varies with transmit signal power, frequency, height, and gain of the antennas. Path loss is directly proportional to d^{-n}, where n is a surrounding environment dependent parameter. This model has been applied numerous times with $n = 1.5 - 1.8$ for LOS [23, 26], $n = 2.4 - 2.8$ for NLOS paths in factory environments [23, 26], $n = 1.81 - 5.22$ for different types of buildings [32, 34], $n = 1.4 - 3.3$ for several manufacturing floors, $n < 2$ for hallway [55, 59] and $n = 3$ for room measurements [85, 86]. A comprehensive table of n values for various scenarios and construction materials is provided in [96]. Lower n is used for free-space propagation and propagation through hallways, rooms, or across parallel walls. Higher n is encountered when transmitted signals penetrate walls, ceilings, floors, etc. or have large number of scatterers across the communication channel. However, this model only provides a predicted mean value of the possible path loss. Hence a standard deviation from the mean value is associated with every set of accumulated measurement.

2.6.2   Model 2

A variable exponent n for the d^{-n} model was proposed in [152]. In this case, n changes with distance d between the transmitter and the receiver. Hence, the received power varies as a function of both distance and environment. For channel measurements in [152], n is allowed to increase from 2 to 12, where $n = 2$ for $1 < d < 10m$, $n = 3$ for $10 < d < 20m$, $n = 6$ for $20 < d < 40m$, and $n = 12$ for $d > 40m$. Measurements are taken with a fixed transmitter located in the middle of a corridor and a mobile receiver moving inside several rooms and corridors on the same and different floors as that of the transmitter. The environment-dependent exponent is considered to increase with the increase in distance between the transmitter and the receiver.

2.6.3   Model 3

Path loss over direct line-of-sight (LOS) link between an ariel node (UAV/flight) and ground node has been modeled and is expressed as

$$P(\text{LOS}) = \prod_{n=0}^{m} \left[1 - e^{-\frac{[h_u - \frac{(n+1/2)(h_u - h_G)}{m+1}]^2}{2\Omega^2}} \right] \quad (2.71)$$

where $m = \text{floor}(r\sqrt{s\xi} - 1)$, r is the horizontal distance between the ariel and the ground nodes, h_u is the vertical height of the ariel node from earth's surface, h_G is the vertical height of the ground node antenna from the earth's surface, s is the ratio of built-up land area to the total land area, ξ is the mean number of buildings per unit area (in Km^2), and Ω is the height distribution of the buildings following Rayleigh distribution, $p(H) = (H/\Omega)^2 - e^{(-H/2\Omega^2)}$.

2.6.4 Model 4

A modified Friis equation has been proposed for calculating path loss in Terahertz (THz) communication channel. The total path loss in that case can be given by

$$PL(dB) = 20 \log \frac{4\pi d}{\lambda_g} + 10\alpha d \log e \quad (2.72)$$

where d is the propagation distance of the wave, $\alpha = 4\pi K/\lambda_0$ is the absorption coefficient with λ_0 as the wavelength of operation, and λ_g is the wavelength experienced over skin depth. The absorption coefficient is a function of the extinction coefficient K, which measures the amount of absorption loss of the electromagnetic wave in the medium.

2.6.5 Model 5

Path loss model 60 GHz communication was first proposed in [202] and the modeling is done based on a fixed transmitter and a fixed receiver. The delay between the transmitter and the receiver is represented as a function of distance between them. Points at fixed distances from the transmitter and the receiver are assumed to be centered in an ellipse in two dimension and an ellipsoid in three dimension. When a wall intersects the ellipsoid, the intersection forms a shape that defines all the points on the wall that are at fixed delay away from the receiver and transmitter. This shape depends on the orientation of the transmitter, receiver, and reflecting walls and is an intersection of an ellipsoid. Hence, the received power for a single reflection at the specified point will be given by

$$P_{point} = \frac{\lambda^2}{(4\pi)^2} \frac{\prod_{i=1} \rho_i}{p^2} \tag{2.73}$$

where p is the path length and ρ_i is the reflection coefficient. For a single point, the reflection coefficient will be ρ_1. The operating wavelength is λ and $p = R_1 + R_2$, where R_1 is the distance from the point to the transmitter and R_2 is that to the receiver. Then we can write

$$P_{point} = \frac{\lambda^2}{(4\pi)^2} \frac{\rho_1}{(R_1 + R_2)^2} \tag{2.74}$$

Measurements are taken inside an empty rectangular room to establish a good fit with the proposed channel model [196].

2.6.6   Model 6

A semi-empirical linear path loss model was proposed in [201] in terms of two types of obstacles, namely glass/wood wall and brick/concrete wall. The path loss is given by

$$L_{lin}(d) = \alpha(d) + l_c(dB) \tag{2.75}$$

where $L_{lin}(d)$ is the linear path loss at distance d from the transmitter, α is the linear coefficient, and l_c is constant. The path loss due to multipath walls can be given by

$$L_{MWL}(d) = \alpha(d) + l_c + \sum_{i=1}^{l} m_i l_i + k_f l_f (dB) \tag{2.76}$$

where m_i is the number of walls of type i, l_i is the loss factor of type i, k_f is the number of floors penetrated in the main path, and l_f is the floor attenuation factor. However, path loss due to moveable obstacles like chairs, tables, and furnitures. have not been considered, while deriving the model.

2.7 Modeling Power Delay Profile

Power Delay Profile (PDP) is defined as the power of the received signal through a wireless communication channel, as a function of time delay. The general form of PDP is given by

$$p(\tau) = \sum_i p_i \delta(\tau - \tau_i) \tag{2.77}$$

where $\sum_i p_i = 1$ is the total power contained in the received signal consisting of i multipath components, $\delta(\tau)$ is the unit impulse response, and τ_i is the delay suffered by the ith multipath component. Identifying the PDP characteristics is an important aspect of wireless channel modeling. Modeling of PDP in wireless environment can be achieved through two kinds of approaches. The impulse response of a signal with bandwidth W can be completely described by a set of samples spaced by $1/W$ or less. Hence, this set of samples ($\tau_i = i/W, i = 0, 1, 2, \dots$) can appropriately characterize the PDP. The second approach is to model the multipath components, arriving with arbitrary delays and phases without any relation with integral multiples of $1/W$. Several PDP models have been proposed over years, each following either of the two above-mentioned approaches. We will be briefly describing some of the models proposed in literature to statistically characterize the PDP in the following subsections.

2.7.1 Exponential Model

The Exponential Model was proposed in [197] based on accumulated measurement from 14 locations within an office building at 500 MHz using baseband pulses. Modeling of PDP is based on discrete samples separated by $1/W$. The first component has a relative amplitude of 1. The successive components decay with time τ, according to the decaying exponential function $r \, exp \, (-\tau/\varepsilon)$, where r is the amplitude ratio of the second component to the first and ε is the decay time constant for all successive components. However, the exponential decay is assumed to be truncated at $\tau = 5\varepsilon$. Hence, the received power of ith component will be given by

$$p_i = \begin{cases} c & \text{if } \tau_i = 0 \\ cr \ exp \ (-\tau_i/\varepsilon) & \text{if } 0 < \tau_i \le 5\varepsilon \end{cases} \qquad (2.78)$$

where $\tau_i = i/W$, c is the normalizing constant appropriately selected so that $\sum_i p_i = 1$. The variables r and ε are assumed to be lognormally distributed over the communication channel such that $10 \ log \ r$ and $10 \ log \ \varepsilon$ are Gaussian random variables. The PDF of the local sample at delay τ is assumed to be Gamma distributed with its corresponding m-parameter having a truncated Gaussian PDF over the transmit–receive path. The mean and variance of the m-parameter are assumed to be a linear decaying function of τ.

2.7.2 Exponential-Lognormal Model

The exponential-lognormal model [199] devises separate ways to characterize PDP in LOS paths and NLOS paths. For the LOS path, the first component arrives with a constant minimum delay, while the subsequent components decay as a function of the product of a decaying exponential and a correlated lognormal random process. For the NLOS path, all components decay exponential-lognormally. Thus, with $\tau_i = i/W$, the received power of the ith component will be given by

$$p_i = \begin{cases} k \ exp \ (-\tilde{\alpha}\tau_i/\bar{\tau}_{rms})s(\tau_i) & \text{if } i \ge 0 \quad \text{NLOS} \\ 10^{A/10} & \text{if } i = 0 \\ k \ exp \ (-\tilde{\alpha}\tau_i/\bar{\tau}_{rms})s(\tau_i) & \text{if } i > 0 \end{cases} \Bigg\} \quad \text{LOS} \quad (2.79)$$

where $\tilde{\alpha}$ is a dimensionless decay constant, A is the direct LOS path amplitude, $s(\tau_i)$ is a correlated lognormal process varying with delay, $\bar{\tau}_{rms}$ is the average rms delay spread and k is the normalizing factor, chosen such that $\sum_i p_i = 1$. The decay constant α varies with distance d between the transmitter and the receiver, according to

$$\alpha = \alpha_0 - \gamma \ log_{10}(d) + \varepsilon \qquad (2.80)$$

where α_0 is a constant, ε is a Gaussian random variable varying from one distance to another with zero mean and standard deviation σ_d and γ is the random variable varying from building to building. The LOS path amplitude A alos varies with d according to

$$A = A_0 - 10\gamma_A \, log_{10}(d) + \varepsilon_A \qquad (2.81)$$

where A_0 and γ_A are constants and ε_A is a Gaussian random variable from one distance to another with zero mean and standard deviation σ_A.

2.7.3   Cluster Model

The Cluster model [86, 198] is based on characterizing multipath components with arbitrary delays and amplitudes and are not uniformly spaced. The PDP will be given by

$$p(\tau) = c \sum_l |\xi_l|^2 \sum_k |\overline{\beta_{k,l}}|^2 \delta(\tau - T_l - \tau_{k,l}) \qquad (2.82)$$

where l is the cluster index ($l = 0, 1, 2, \dots$), k is the ray index within a cluster ($k = 0, 1, 2, \dots$), T_l is the delay of the first ray of the lth cluster, $\tau_{k,l}$ is the delay of the kth ray of the lth cluster, $|\overline{\beta_{k,l}}|^2$ is the locally average power of the kth ray of the lth cluster and c is the normalizing factor, chosen such that $\sum_i p_i = 1$. The cluster scale factor $|\xi_l|^2$ is assumed to be lognormally distributed random variable, varying from cluster to cluster and path to path. The cluster delay, T_l and the ray delay within a cluster, $\tau_{k,l}$ are both assumed to be Poisson distributed, with average arrival rates Λ and λ, respectively. The average ray amplitude $|\overline{\beta_{k,l}}|^2$ is given by

$$|\overline{\beta_{k,l}}|^2 \propto exp\left(-T_l/\Gamma\right) exp\left(-\tau_{k,l}/\gamma\right) \qquad (2.83)$$

where the cluster amplitudes decay exponentially with time constant Γ and the rays within a cluster will decay exponentially with time constant γ. Clustering-based model has not been restricted to only modeling PDP but has been extended to model radio channel, which is now a widely established technique for channel modeling.

2.7.4   Tapped Delay Line Model

Uniformly distributed tapped delay line model for characterizing PDP [195, 196, 197] is based on mean power measurements of large number of received impulse responses (IRs) in wireless environment. Only significant paths are considered for modeling and hence paths with average power below a certain level are discarded. Let us assume K normalized IRs each sampled by N complex values, equally spaced by $\Delta\tau$ time interval. Then, the average PDP can be calculated as

$$p(n\Delta\tau) = \frac{1}{K} \sum_{k=0}^{K-1} |h(n\Delta\tau, k\Delta t + t_1)|^2 \qquad 0 \le n \le N - 1 \quad (2.84)$$

where $k\Delta t + t_1$ is fixed observation instant of kth IR and h is normalized IR. Paths with average power below $-20dB$ will be discarded. PDP with sample amplitudes $[-20, 0]$ dB is shorter than N. Let L' be the number of samples in those paths such that $L' \le N$. Let us also assume that we want to get M tapped lines with delays τ_m such that L' samples will be grouped into M groups. If $L'/M = I'$ is not an integer, samples with magnitude zero are added with an integer number I. The tapped delay τ_m will then be given by,

$$\tau_m = \frac{\sum_{n=i_m}^{i_m+I-1} n\Delta\tau . P(n\Delta\tau)}{\sum_{n=i_m}^{i_m+I-1} P(n\Delta\tau)} \qquad 0 \le m \le M - 1 \quad (2.85)$$

where i_m is the sample number at the start of each group. The complex amplitude over interval m for the kth IR at a delay τ_m and fixed observation instant $k\Delta t + t_1$ will be given by,

$$h_m(\tau_m, k\Delta t + t_1) = \frac{1}{I} \sum_{n=i_m}^{i_m+I-1} h(n\Delta\tau, k\Delta t + t_1) \qquad (2.86)$$

For K IRs and every τ_m, we can get K complex tap amplitude values, which can provide the statistical distribution for each tap amplitude. The tap amplitude at τ_m is the average value of K amplitude values.

2.7.5   ACT technology-based FIR modeling

Finite Impulse Response (FIR) models have long been applied to the study of multipath propagation channels. Emulation of complex path amplitude and delay based on Acoustic Charge Transport (ACT) technology was first proposed in [194]. It follows the line of FIR models using ACT Programmable Transversal Filter (PTF). The filter taps are uniformly distributed and programmed to real amplitudes. The complete IR model is implemented by considering the operating bandwidth and convolving a complex bandpass kernel with the complex theoretical IR ray model. The IR ray model is based on the desired statistical or deterministic model used to derive path delays and amplitudes. Such a bandpass kernel could take the form of $h_k(t)$ where,

$$h_k(t) = Sa(\pi W t)e^{j2\pi f_0 t} \tag{2.87}$$

where W is the operating bandwidth, f_0 is the center frequency, and $Sa(\theta) = \sin(\theta)/\theta$.

2.8   Modeling Frequency Dependence of Channel Statistics

The frequency-dependent impulse response model of the mobile radio channel can characterize the mobile channel at different frequencies as

$$h(t, f_i) = \sum_{n=1}^{N(f_i)} \alpha_n(f_i)\delta(t-\tau_n(f_i))e^{j\theta(f_i)} \qquad i = 0, 1, 2, \ldots, I \tag{2.88}$$

where $h(t, f_i)$ is the impulse response of the channel at time t and i^{th}-operating frequency (f_i), δ is the Dirac delta function, and I is the number of frequencies to he investigated. The parameters amplitude α_n, delay τ_n, phase θ_n, and number of multipath components N are functions of frequency (f_i). The main idea of frequency dependent modeling is that based on the knowledge of channel's parameters at the basic frequency (f_0), and incorporating the assumed correlations of channel parameters at the frequency (f_i) with the frequency (f_0), the channel's parameters at other frequency ($f_i = f_0 + \Delta f_i$) will be predicted.

2.8.1 Amplitude modeling

In mobile radio channels, the Rayleigh distribution is a well and widely accepted model to describe the amplitude fluctuations as

$$f_A(\alpha_n) = \frac{\alpha_n}{\sigma_n^2} \, exp \, (-\frac{\alpha_n^2}{2\sigma_n^2}) \qquad \alpha_n \geq 0 \qquad (2.89)$$

where σ_n is a Rayleigh parameter.

2.8.2 Number of multipath components modeling

The number of multipath components arriving during an interval of maximum excess delay T_{max} is equal to $N = \lambda * T_{max}$, where λ is the mean arrival rate of multipath components. A model for the arrival rate of multipath components as function of frequency is introduced as

$$\lambda(\Delta f_i) = \left\{ \begin{array}{l} {}_0 + \lambda_{max}[1 - \, exp \, (-\gamma_H \Delta f_i)] \\ {}_0 + \lambda_{min}[1 - \, exp \, (-\gamma_L \Delta f_i)] \end{array} \right. \qquad (2.90)$$

where λ_0 is the arrival rate of the multipath components at basic frequency, λ_{min} and λ_{max} are the minimum and the maximum arrival rates to be achieved, respectively, and γ_H and γ_L are two arbitrary constants.

2.8.3 Arrival time modeling

If the objects causing the multipath (reflection and refraction) are located randomly throughout the space surrounding the link, then path arrival times form a Poisson sequence. The probability of N arrivals during the interval T can be given by

$$\text{Prob} \, [N \text{ arrivals in } T_{(sec)}] = \frac{(\lambda T)^N}{N!} e^{\lambda T} \qquad (2.91)$$

where λ is the mean arrival rate (Poisson parameter), which can be a function of frequency.

2.8.4 Modeling of Frequency Correlation

Correlation between parameters at two frequencies (or time delays) decays with frequency (time). Thus, we propose a correlation model where the correlation coefficient between amplitudes decreases with increasing of the frequency (excess delay). Jake's model can be used which can be written as

$$\rho(x) = J_0^2(\kappa.x) \qquad (2.92)$$

where x can be frequency or excess delay, $\rho(\cdot)$ is the correlation coefficient, $J_0(\cdot)$ is the Bessel function of first kind and order zero, and κ is a parameter controlling the rate of correlation decrease with x.

To model the frequency correlation by the presence of the excess delay correlation, two methods have been developed in [193]. In Method I, it is assumed that, for each generated impulse response collected at each frequency, the LOS component is present. Taking LOS as reference, first the correlation as function of the excess delay is performed for the basic frequency. Then, the amplitude correlation between two LOS components at two frequencies is applied according to (87). Next, the same procedure of the amplitude correlation as function of the excess delay is repeated for the frequency under study. In method II the presence of LOS is not necessary. By taking the strongest path component (SFC) as reference, the amplitude correlation as function of excess delay is performed for the basic frequency. Then, the amplitude correlation as function of frequency is preformed between the components, which arrive at the same moment of the excess delay axis (i.e., the same bin index) and at different frequencies.

2.8.5 Modeling of RMS Delay Spread

The rms delay spread τ_{rms} limits the maximum data transmission rate, which can reliably be supported by the channel. In the case without using diversity or equalization, the τ_{rms} is inversely proportional to the maximum usable data rate of the channel [200]. The correlation between rms delay spreads at different frequencies is given as

$$\rho_{\tau_{rms}}(\Delta f_i) = E\{[\tau_{rms}(f_0 + \Delta f_i) - E(\tau_{rms}(f_0 + \Delta f_i))] \cdot [\tau_{rms}(f_0) \\ - E(\tau_{rms}(f_0))]\}\{\sqrt{Var[\tau_{rms}(f_0 + \Delta f_i)]Var[\tau_{rms}(f_0)]}\}^{-1} \\ (2.93)$$

The τ_{rms} for two impulse response estimates at two frequencies have a bivariate normal distribution with their correlation coefficient a decreasing function of frequency spacing. The bivariate function is given by

$$f_{X_{f_i},X_{f_0}}(x_{f_i}, x_{f_0}) = \frac{1}{2\pi\sigma_{f_i}\sigma_{f_0}\sqrt{1-\rho^2}} \, exp \left\{ - \frac{1}{2(1-\rho^2)} \right. \\ \left. \left[\frac{(x_{f_i} - \mu_{f_i})^2}{\sigma_{f_i}^2} - 2\rho\frac{(x_{f_i} - \mu_{f_i})(x_{f_0} - \mu_{f_0})}{\sigma_{f_i}\sigma_{f_0}} + \frac{(x_{f_0} - \mu_{f_0})^2}{\sigma_{f_0}^2} \right] \right\} \\ (2.94)$$

where X_{f_0} and X_{f_i}, represent the τ_{rms} values for the impulse response profiles collected at the basic frequency and the frequency under study, respectively, $\mu_{f_0} = E[X_{f_0}]$, $\mu_{f_i} = E[X_{f_i}]$, $\sigma_{f_0}^2 = Var[X_{f_0}]$, $\sigma_{f_i}^2 = Var[X_{f_i}]$ and ρ is the correlation coefficient between X_{f_0} and X_{f_i}.

2.9   Hybrid Models

In this section, we will be summarizing some radio channel models that have been proposed over literature for different application and propagation scenarios. They have been deduced from a combination of different approaches explained in previous sections. Each of these models has proved to be stand-alone modeling approach for a particular scenario or application, based on multiple measurement campaigns followed by model extraction, simulation and analysis.

2.9.1   3-D Stochastic Image-based Model

A 3-D stochastic image-based channel (SIBIC) model for UWB 3-D sensor arrays was proposed in [183]. This model treats the virtual point sources as clusters of multipath components. A point-source transmitter is assumed, whose image is found using the image method. The inter-cluster arrival times are extracted from the time of arrival (TOA) of the images and mean angle of arrival (AOA) of each cluster is calculated from AOA of each image (refer to Fig. 2.2). Then the intra-cluster MPC arrival times and AOA of each MPC in each cluster are generated by a Poisson process and a Laplacian distribution function, respectively. Non-integer sampling and receiver bandwidth effects are also taken into account in SIBIC. The SIBIC model is based on five major components.

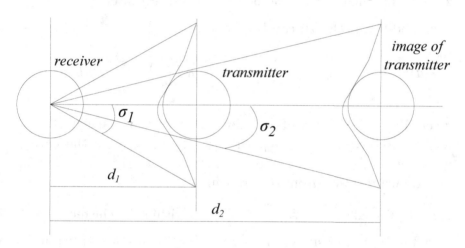

FIGURE 2.2

Angular speed of transmitter and its image as perceived by receiver

2.9.1.1   Calculation of the position of the cluster center

The position vectors of all virtual sources can be given by

$$R_r + R'_p = (x' - 2qx' + 2oL_x,\ y' - 2jy' + 2lL_y,\ z' - 2tz' + 2mL_z) \quad (2.95)$$

where $X' = (x', y', z')$ represents the point source transmitter, $X = (x, y, z)$ represents the receiver, $R'_p = (x' - 2qx',\ y' - 2jy',\ z' - 2tz')$, where eight permutations of vector $R'_p = (\pm x', \pm y', \pm z')$ are calculated as function of $p = (q, j, t)q, j$ and $t \in \{0, 1\}$. The translation vector is given by $R_r = 2(oL_x, lL_y, mL_z)$ where $o, l, m \in \mathbb{Z} \geq 0$ and L_x, L_y, L_z are dimensions of the rectangular room with o, l and m varying over a specified range. The distance between each virtual source and the receiver is given by

$$d = |R_r + R'_p - X| \qquad (2.96)$$

and the AOA information vector k is given by

$$k = \frac{R_r + R'_p - X}{|R_r + R'_p - X|} \qquad (2.97)$$

2.9.1.2   Calculation of path loss of the cluster center

The modified path loss can be given by

$$P_L(d) = -20 \, log_{10} \left(\frac{4\pi d_0}{\lambda_c}\right) - 10n \, log_{10} \left(\frac{d}{d_0}\right) + \sum_{i=1}^{W} 20 \, log_{10} (\Gamma_i) \qquad (2.98)$$

where W is the number of partitions between the image at $R_r + R'_p$ and the receiver, n is the path loss exponent, λ_c is the carrier wavelength, Γ_i is the reflection coefficient at the ith plane, and d_0 is a known distance from the transmitter.

2.9.1.3   Calculation of AOA of the rays within the cluster

The 3-D angular distribution is given by the product of two independent Laplacian functions given by

$$p(\theta, \phi) = \frac{1}{\sqrt{2}\sigma} e^{-|\sqrt{2}\theta/\sigma|} \frac{1}{\sqrt{2}\sigma} e^{-|\sqrt{2}\phi/\sigma|} \qquad (2.99)$$

where θ denotes the zenith angle, ϕ denotes the azimuth angle relative to the cluster mean, and σ is the angular standard deviation. If at distances d_1 and d_2, the angular spreads of a cluster are σ_1 and σ_2, respectively, the relationship between σ_1 and σ_2 will be

$$\sigma_2 = \tan^{-1} \left(\frac{d_1 \tan^{-1}(\sigma_1)}{d_2}\right) \qquad (2.100)$$

2.9.1.4   Calculation of TOA of the rays within the cluster

The received multipath signals are assumed to be clustered whose arrival times are modeled by a double Poisson distribution function, one part models the intra-arrival times and the other part models the inter-arrival times.

$$p(t_{k+1}|t_k) = \lambda e^{-\lambda(t_{k+1}-t_k)} \tag{2.101}$$

where λ is the intracluster rate, T_l is the arrival time of the lth cluster, $\beta(T_l)$ is the amplitude of the first ray of the cluster, and t_k and t_{k+1} are the TOAs of the kth and the $(k+1)^{st}$ rays, respectively. The amplitude of the $(k+1)^{st}$ ray in the cluster can be given by

$$\beta(T_l, k+1) = \beta(T_l)e^{-(t_{k+1})/\gamma} \tag{2.102}$$

where γ is the exponential power delay constant of the cluster.

2.9.1.5   Calculating the receiver effect

Considering non-integer sampling rate and finite receiver bandwidth, the channel response can be given by

$$h(t) = \begin{cases} \frac{1}{2}[1 + \cos(2\pi t/T_w)](2\pi f_c t) & -T_w/2 < t < T_w/2 \\ 0 & \text{otherwise} \end{cases} \tag{2.103}$$

where T_w is the Hanning window duration and f_c is the receiver bandwidth.

2.9.2   Auto-regressive Model

Direct application of statistical auto-regressive modeling technique for radio propagation has been proposed in [109]. In this model, first of all mean and standard deviation of AWGN is calculated from the known noise level on the tail of an IR profile. Next, a power level of 4 standard deviations above the mean noise power level is assigned. Finally the rms delay, τ_{rms} is calculated ignoring all information below the noise reference level. The measured transfer function of a radio channel at each location can be expressed as the output of an auto-regressive process of order p. The auto-regressive (AR) process can then be given by

$$H(f_n, x) - \sum_{k=1}^{p} a_k H(f_{n-k}, x) = \nu(f_n) \qquad (2.104)$$

where $H(f_n, x)$ is the nth sample of the transfer function at location x, a_k is a complex model parameter and $\nu(f_n)$ is the complex AWGN. The Z-transform of $H(f_n, x)$ will be expressed as

$$G(z) = \prod_{k=1}^{p} \frac{1}{1 - p_k z^{-1}} \qquad (2.105)$$

The AR process $H(f_n, x)$ therefore represents the radio propagation channel and is characterized by the location of p poles. Other parameters like coherence bandwidth, rms delay, and path loss can be calculated from this model for complete identification of the communication environment.

2.9.3 Two Ring Model

A two-ring model was proposed in [192], where the scatterers local to transmitter and receiver are modeled to be distributed on two separate rings. A single-bounce scattering is assumed while formulating this model. Over the forward transmission channel, the single bounce rays from scatterer S_i around the receiver and scatterer S_k' around the transmitter, arrive at the receiver. Let D be the distance between the receiver and the transmitter. The radii of the rings of scatterers around the transmitter and the receiver are assumed to be R and R', respectively. The directions of transmit and receive arrays are denoted by α_{pq} and β_{lm}. For a frequency non-selective communication link between the element BS_p and the element MS_l, the complex low pass channel gain $h_{lp}(t)$ will be given by

$$h_{lp}(t) = \sqrt{\eta' \Omega_{lp}} \lim_{N' \to \infty} \frac{1}{\sqrt{N'}} \sum_{k=1}^{N} {}' g_k'$$

$$\times exp\left\{ j\psi_k' \frac{j2\pi}{\lambda} (\xi_{pk}' + \xi_{kl}') + j2_d \cos(\varphi_k' - \gamma)t \right\}$$

$$+ \sqrt{\eta \Omega_{lp}} \lim_{N \to \infty} \frac{1}{\sqrt{N}} \sum_{i=1}^{N} g_i$$

$$\times exp\left\{ j\psi_i \frac{j2\pi}{\lambda} (\xi_{pi} + \xi_{il}) + j2_d \cos(\phi_i - \gamma)t \right\} \qquad (2.106)$$

where Ω_{lp} is the transmitted power, $\Omega_{lp} = E[|h_{lp}|^2] \le 1$, η' and η represent the contributions of transmit and receive rings such that $\eta' + \eta = 1$, N and N' are the number of independent scatterers around transmitter and receiver, respectively, g_i and g_k' are the amplitudes of the waves scattered by the S_i and S_k', respectively, ψ_i and ψ_k' are the associated phase shifts, *varphi*$_i$ and ϕ_k' are the AOD of the waves travelling from the transmitter and, ϕ_i and φ_k' are the AOA of the waves toward the receiver. The sets $\{g_i\}_{i=1}^{\infty}$ and $\{g_k'\}_{k=1}^{\infty}$ consist of positive independent random variables with finite variances, independent of $\{\psi_i\}_{i=1}^{\infty}$ and $\{\psi_k'\}_{k=1}^{\infty}$. It is assumed that $\{\psi_i\}_{i=1}^{\infty}$ and $\{psi_k'\}_{k=1}^{\infty}$ are uniform and i.i.d. random variables with uniform distributions over $[0, 2\pi)$. Setting $N^{-1} \sum_{i=1}^{N} E[g_i^2] = 1$ and $N'^{-1} \sum_{k=1}^{N} {}' E[g_k'^2] = 1$, the other characterizing parameters for channel modeling can be derived both for SISO and MIMO systems.

2.9.4 Multiple Scatterer Model

A new statistical model for characterizing time variations of both LOS and NLOS fixed indoor communications (FIC) channels is proposed in [190, 191]. Let N be the number of scatterers, each moving with a non-negative speed, $v^{(i)}$ m/s for $i \in 1, 2, \ldots N$. Both transmitted and received signals are assumed to be vertically polarized (refer to Fig. 2.9.4). The total received signal will be given by

$$r(t) = \sum_{i=0}^{N} r^{(i)}(t) = Re[\tilde{r}(t)\ exp\ (j2\pi f_c t)] \qquad (2.107)$$

where $r^{(0)}(t)$ is the LOS component and $r^{(i)}(t)$ is the received

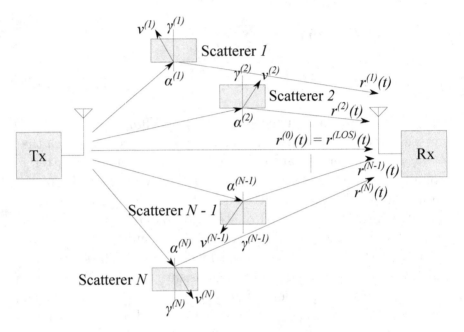

FIGURE 2.3
Indoor radio channel with multiple scatterers

component due to scatterer i, f_c is the carrier frequency and $\tilde{r}(t)$
is the total received complex envelope given by

$$\tilde{r}(t) = \sum_{i=0}^{N} r^{(i)}(t) = \sum_{i=0}^{N} \sum_{l=1}^{M^{(i)}} C_l^{(i)} e^{-j\phi_l^{(i)}(t)} \tilde{s}(t - \tau_l^{(i)}) \qquad (2.108)$$

where $\tilde{s}(t)$ is the transmitted complex envelope, $M^{(i)} \in 1, 2, \ldots$ are
the number of scattered waves in $r^{(i)}(t)$, $C_l^{(i)}$ and $\phi_l^{(i)}(t)$ denote the
magnitude and the phase shift of the lth scattered wave of $r^{(i)}(t)$,
respectively. Let us assume that $\tilde{s}(t) = 1$. Then the time variation

of an FIC channel can be identified by the time variation of $r(t)$, which can be expressed as

$$r(t) = \underbrace{\left(\sum_{i=0}^{N} \sum_{l=1}^{M^{(i)}} C_l^{(i)} \cos \phi_l^{(i)}(t) \right) \cos 2\pi f_c t}_{r_I(t)}$$

$$\underbrace{- \left(\sum_{i=0}^{N} \sum_{l=1}^{M^{(i)}} C_l^{(i)} \sin \phi_l^{(i)}(t) \right) \sin 2\pi f_c t}_{r_Q(t)} \qquad (2.109)$$

where $r_I(t)$ and $r_Q(t)$ are the in-phase and quadrature components of $r(t)$, respectively. If $\theta_l^{(i)}$ denote the received AOA of the lth scattered wave, $\alpha_l^{(i)}$ denote the signal scattering angle of the lth wave and $\gamma_l^{(i)}$ is the angle between i^{th} scatterer's direction of movement and the direction perpendicular to the reflecting surface of the ith scatterer, the time variation of $r(t)$ can be captured by its auto-correlation functions (ACF) as

$$E[r(t)r(t + \Delta t)] = E_{\vec{\tau}, \vec{\alpha}, \vec{\gamma}, \vec{\theta}} \left[r_I(t) r_I(t + \Delta t) \right] \cos 2\pi f_c \Delta t$$

$$- E_{\vec{\tau}, \vec{\alpha}, \vec{\gamma}, \vec{\theta}} \left[r_Q(t) r_Q(t + \Delta t) \right] \sin 2\pi f_c \Delta t \quad (2.110)$$

where $\vec{\tau} = [\tau_1^{(0)} \dots \tau_{M(N)}^{(N)}]^T$ denotes the path delay, $\vec{\alpha} = [\alpha_1^{(0)} \dots \alpha_{M(N)}^{(N)}]^T$, $\vec{\gamma} = [\gamma_1^{(0)} \dots \gamma_{M(N)}^{(N)}]^T$ and $\vec{\theta} = [\theta_1^{(0)} \dots \theta_{M(N)}^{(N)}]^T$.

2.9.5 Modified Saleh-Valenzuela Model

Based upon clustering effect of multipath components, an UWB channel model is proposed in [188, 189] based on Saleh-Valenzuela model. The CIR can be given by

$$h(t) = \sum_{l=0}^{L} \sum_{h=0}^{K_l} \alpha_{k,l} \delta(t - T_l - \tau_{k,l}) \qquad (2.111)$$

where L is the number of clusters, K_l is the number of MPCs within the lth cluster, $\alpha_{k,l}$ is the multipath gain coefficient of the

k^{th} component in the lth cluster, T_l is the TOA of the first arriving MPC of the lth cluster and $\tau_{k,l}$ is the delay of the kth MPC relative to the lth cluster arrival time. The proposed channel model is based on inter-cluster parameter L, T_l and intra-cluster parameters $K_l, \tau_{k,l}, \alpha_{k,l}$. the ray arrival times are given by a combination of two Poisson processes,

$$p(\tau_{k,l}|\tau_{(k-1),l}) = \beta\lambda_1 \; e^{[-\lambda_1(\tau_{k,l}-\tau_{(k-1),l})]} + (\beta - 1)\lambda_2$$
$$\times \; e^{[-\lambda_2(\tau_{k,l}-\tau_{(k-1),l})]} k > 0 \qquad (2.112)$$

where β is the mixture probability and, λ_1 and λ_2 are the ray arrival rates. The average power of a MPC at a given delay, $T_k + \tau_{k,l}$ is given by

$$\overline{\alpha_{k,l}^2} = \overline{\alpha_{0,0}^2} \cdot e^{-T_l/\Gamma} \cdot e^{-\tau_{k,l}/\gamma} \qquad (2.113)$$

where $\overline{\alpha_{0,0}^2}$ is the expected power of the first arriving MPC, Γ is the decay exponent of the cluster and γ is the decay constant of the rays within a cluster. The temporal correlation between adjacent power amplitudes can be defined by the amplitude temporal correlation coefficient, $\rho_{\alpha_k,k+1}$ is given by

$$\rho_{\alpha_k,k+1} = \frac{E\{(\alpha_k - \overline{\alpha_k})(\alpha_{k+1} - \overline{\alpha_{k+1}})\}}{\sqrt{E\{(\alpha_k - \overline{\alpha_k})^2(\alpha_{k+1} - \overline{\alpha_{k+1}})^2\}}} \qquad (2.114)$$

where α_k and α_{k+1} are the power amplitudes of the kth and the $(k+1)^{th}$ components, respectively. However, for a wireless environment, $\rho_{\alpha_k,k+1}$ exhibits low values due to low correlation between different MPCs from different scatterers.

2.9.6 Cluttering Model

An efficient hybrid model combining a two-dimensional (2-D) site-specific model and a statistical model characterizing multipath fading was proposed in [187] for radio propagation environment. This model can predict mean field strength due to diffused scattering with a factor r, which is defined as the measure of the cluttering strength of the propagation environment. The total received field strength is given by

$$E_r = E_{dr} + E_{sr} \qquad (2.115)$$

where $E_{dr} = \vec{\rho^*} \cdot \sum_i \vec{E}_{d_i}$ denotes the deterministic ray field, $E_{sr} = \vec{\rho^*} \cdot \sum_i \vec{E}_{s_i}$ denotes the random ray field and ρ^* is the complex conjugate of unit polarization vector of the receive antenna. The 2-D site-specific model will therefore be given by

$$E_{dr} = \sum_i E_{1m} G_{ti} G_{ri} L_i(d) \prod_i R_c(\theta_{ji}) \prod_m T(\theta_{mi}) \qquad (2.116)$$

where E_{1m} is the field envelope 1 m away from the transmitting antenna; G_{ti} and G_{ri} are the field-amplitude radiation patterns of the transmitting and the receiving antennas, respectively; $L_i(d)$ is the free-space path loss of the th ray with an unfold length d; R_c and T are the reflection and transmission coefficients, respectively; and θ_{ji} and θ_{mi} are the jth and m^{th} reflecting and transmitting angles, respectively. The received scattered field will be given by

$$E_{sr} = E_0 \sum_{n=1}^{N} exp\left(j\phi_n\right) \qquad (2.117)$$

where ϕ_n is a random phase that is uniformly distributed throughout 0 to 2π ; N is the total number of received rays; and E_0 is an arbitrary constant. Finally the spatial correlation between consecutive scattered field is given by

$$R_{E_{sr}}(d) = \langle E_{sr}(r) E_{sr}^*(r+d) \rangle = N E_0^2 J_o(kd) \qquad (2.118)$$

where $\langle \cdot \rangle$ and J_o are the ensemble average and the zero-order Bessel function of the first kind, respectively; and k and d are the free-space wavenumber and spatial distance, respectively.

2.9.7 Reduced Finite Difference Time Domain (RFDTD) Model

A novel memory-efficient formulation of Finite Difference Time Domain (FDTD), used to predict the statistics for the wireless channel for a residence, has been proposed in [186]. The basic model used for the time response of the baseband wireless channel is given by

$$c(t, x, y) = \frac{1}{d^{\beta/2}} \sum_k \alpha_k(x, y)\delta(t - k\tau) \qquad (2.119)$$

where d is the distance between the transmitter and the oberva-
tion point (x, y), $\alpha_k(x, y)$ are the complex coefficients of individual multipath components and τ is the time resolution used. The
large-scale path loss exponent β is determined by a linear fit on
the logarithm of the average power of channel profiles for several
points inside the residence. Then the location-dependent rms delay
spread is calculated from

$$\tau_{rms} = \sqrt{\left\{ \frac{\sum_k (k\tau - \tau_m(x,y))^2 |\alpha_k(x,y)|^2}{\sum_k |\alpha_k(x,y)|^2} \right\}} \qquad (2.120)$$

where $\tau_m(x, y)$ is the mean excess delay. The distributions of the
magnitudes of α_0 and α_1 can also be computed using R-FDTD and
compared to the ones obtained using measured data. Lognormal
distributions are used for the fitting of experimental and simulated
data. Since lower values of τ_{rms} is observed, the channel can be
modeled with a single complex tap at the frequency of interest by
simply taking the Fourier transform of (115).

$$C(f_0, x, y) = \sum_k \alpha_k(x, y) \, exp \, (-j2k\pi f_0\tau) \qquad (2.121)$$

The complex auto-correlation coefficient is calculated by

$$\rho_C(\Delta x, \Delta y) = \frac{E[C(f_0, x, y)C^*(f_0, x + \Delta x, y + \Delta y)]}{\sqrt{E[|C(f_0, x, y)|^2]}} \qquad (2.122)$$

where $*$ denotes complex conjugate. Rayleigh and Nakagami dis-
tributions are used for the fitting of the distribution of $|C(f_0, x, y)|$.

2.9.8   Wavelet Packet-based Model

A method to model wireless communication channels using wavelet
packets as bases is proposed in [185]. Wavelet packets offer the advantage of having better orthogonality properties and capability
to effectively suppress certain types of noise/interference. Wireless
communication channels can be fully described by their complex
IRs, and wavelet packet-based model only use the magnitude information of CIRs (refer to Fig. 2.9.8). For a time-invariant channel, the CIR $h(t)$ can be represented by a set of wavelet packets

$P_j(t), j = 1, 2, \ldots, N$ as

$$h(t) = \sum_{j=1}^{N} A_j P_j(t) \tag{2.123}$$

A time-varying CIR $h(t, \tau)$ can be expanded (modeled) by two sets of wavelet packets as

$$h(t, \tau) = \sum_{j=1}^{D_i} \sum_{i=1}^{D_0} A_{ij} P_j(\tau) Q_i(t) \tag{2.124}$$

where P and Q are two sets of wavelet packet bases and $D_i \times D_0$ constant channel matrix A characterizes the channel. With the WP-based modeling, the received signal will be given by

$$y(t) = \sum_{i=1}^{D_0} Q_i(t) \sum_{j=1}^{D_i} A_{ij} (P_j * x)(t) \tag{2.125}$$

where $*$ means convolution operation and $x(t)$ is the transmitted signal.

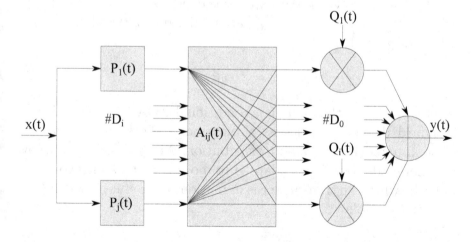

FIGURE 2.4
Wavelet Packet-based time-varying channel model

2.9.9 M-Step 4-State Markov Channel Model

A novel stochastic wide-band dynamic spatio-temporal channel model, which incorporates both the spatial and temporal domain properties as well as the dynamic evolution of paths when the mobile moves is proposed in [184]. A new stochastic channel model based on the Markov process is formulated, that will take into account the time-varying properties of the channel by incorporating the dynamic evolution of paths, when the transmitter is in motion. The spatio-temporal variations of paths within their lifespans are modeled by a Gaussian PDF. The power variations are modeled by a simple low-pass filter (LPF) (refer to Fig. 2.9.9). The dynamic directional channel model can be characterized by the distance-variant directional CIR as

$$h(n; \tau, \phi) = \sum_{l=1}^{L_T(n)} \alpha_l(n) \cdot \delta[\tau - tau_l(n), \phi - \phi_l(n)] \qquad (2.126)$$

for $n = 1, \dots, N$, where $L_T(n)$ is the total number of active paths in the nth fast Doppler block (FDB), while $\alpha_l(n)$, $\tau_l(n)$, and $\phi_l(n)$ are the complex path gain, TOA, and AOA of the lth path in the nth FDB, respectively. An exponential pdf is used to model L_T, while $\{\tau_l\}_{l \in L_T(n)}$ and $\{\phi_l\}_{l \in L_T(n)}$ are modeled by a two-dimensional joint pdf given by

$$f(\tau_l, \phi_l) = f(\phi_l | \tau_l) \cdot f(\tau_l) \qquad (2.127)$$

where $f(\phi_l | \tau_l)$ is the Gaussian conditional AOA pdf in which the standard deviation varies as a function of TOA, which can be approximated by a Weibull distribution. Next a 4-state Markov channel model (MCM) is adapted to model the dynamic evolution of paths when the transmitter is in motion. Each state of the MCM is defined as follows:

- S_0 - no "birth" or "death" $(B_0 D_0)$;

- S_1 - one "death" only $(B_0 D_1)$;

- S_2 - one "birth" only $(B_1 D_0)$;

- S_3 - one "birth" and one "death" $(B_1 D_1)$

The probabilistic switching process between states in the channel model is controlled by the state transition probability matrix **P** given by

$$\mathbf{P} = p_{i,j} = \begin{pmatrix} p_{00} & p_{01} & p_{02} & p_{03} \\ p_{10} & p_{11} & p_{12} & p_{13} \\ p_{20} & p_{21} & p_{22} & p_{23} \\ p_{30} & p_{31} & p_{32} & p_{33} \end{pmatrix} \qquad (2.128)$$

where i and j denote the state index, while $p_{i,j}$ is the state probability that a process currently in state i will occupy state j after its next transition such that $p_{i,j}$ must satisfy

$$0 \leq p_{i,j} \leq 1, \qquad 1 \leq i,j \leq K-1 \qquad (2.129)$$

$$\sum_{j=0}^{K-1} p_{i,j} = 1, \qquad i = 0,1,\ldots,K-1 \qquad (2.130)$$

where K is the number of states and **P** characterizes how paths appear and disappear when the transmitter moves.

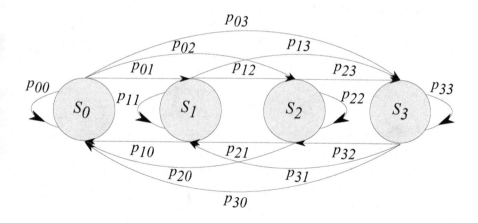

FIGURE 2.5
State transition diagram of the 4-state MCM

The total distance traveled by the MT can be divided into N FDBs. In order to account for the multiple events case, a multiple step (M-step) MCM is proposed. Hence, by applying the M-step 4-state MCM, both the correlation between L_B (birth of paths) and L_D (death of paths) and the multiple births and deaths will be taken into consideration. The number of steps M is determined by the maximum number of births, $L_{B,Max}$, and maximum number of deaths, $L_{D,Max}$, for a particular measurement file. The Birth–Death correlation coefficient, ρ_{BD}, between L_B and L_D is calculated for each measurement file and is defined by,

$$\rho_{BD} = \left| \frac{\sum_{n=1}^{N}(L_B(n) - \bar{L}_B)(L_D(n) - \bar{L}_D)}{\sqrt{\sum_{n=1}^{N}(L_B(n) - \bar{L}_B)^2 \sum_{n=1}^{N}(L_D(n) - \bar{L}_D)^2}} \right|, \quad (2.131)$$

$$\rho_{BD} \in [0,1] \quad (2.132)$$

where $L_B(n)$ and $L_D(n)$ are the number of births and deaths of paths in the nth FDB, respectively, while $\bar{L}_B$ and $\bar{L}_D$ are the mean number of births and deaths, respectively.

Bibliography

[1] L. P. Rice, "Radio transmission into buildings at 35 and 150 mc," in *Bell Syst. Tech. J.*, vol. 38, no 1, pp. 197-210, Jan. 1959.

[2] H. H. Hoffman and D. C. Cox, "Attenuation of 900 MHz radio waves propagating into a metal building," in *IEEE Trans. Antennas Propagat.*, vol. AP-30, no.4, pp. 808-811, July 1982.

[3] D. C. Cox, R. R. Murray, and A. W. Noms, "Measurements of 800 MHz radio transmission into buildings with metallic walls," in *Bell Syst. Tech. J.*, vol. 62, 110.9, pp. 2695-2717, Nov. 1983

[4] D. C. Cox, R. R. Murray, A. W. Noms, "800-MHz attenuation measured in and around suburban houses," in *ATT Bell Lab. Tech. J.*, vol. 63, no.6, pp. 921-954, July-Aug. 1984.

[5] D. C. Cox, R. R. Murray, A. W. Noms, "Antenna height dependence of 800 MHz attenuation measured in houses," *IEEE Trans. Vehicular Techn.*, vol. VT-34, no.2, pp. 108-115, May 1985.

[6] P. J. Barry and A. G. Williamson, "Radiowave propagation into and within a building at 927 MHz," *Electn. Letters*, vol. 23, no.5, pp. 248-249, Feb. 1987

[7] P. J. Barry and A. G. Williamson, "UHF radiowave signal propagation into and within buildings," *in Digest IREECON '87 Conf.* Sydney, Sept. 14-18, 1987, pp. 646-649.

[8] P. J. Barry and A. G. Williamson, "Modeling of UHF radiowave signals within extremally illuminated multi-story

buildings," *J. IERE*, vol. 57, no.6 (Supplement), pp. S231-S240, Nov./Dec. 1987.

[9] A. M. D. Turkamani, J. D. Parsons, and D. G. Lewis, "Radio propagation into buildings at 441, 900 and 1400 MHz," *in Fourth Int. Conf. Land Mobile Radio*, Univ. of Warwick, Conventry, Publication 78, Dec. 1987, pp. 129-139.

[10] A. M. D. Turkamani, J. D. Parsons, and D. G. Lewis, "Measurement of building penetration loss on radio signals at 441, 900, and 1400 MHz," *J. IERE*, vol 58, no.6 (Supplement), pp. 169-174, 1988.

[11] A. F. Toledo and A. M. D. Turkmani, "Propagation into and within buildings at 900, 1800, and 2300 MHz," *in Proc. IEEE Vehicular Techn. Conf. VTC '92*, Denver, Colo., May 1992, pp. 633-636.

[12] S. J. Patsiokas. B. K. Johnson, J. L. Dailing, "The effects of buildings on the propagation of radio frequency signals," *in Proc. of Infr. Conf. Commun., ICC'87*, pp. 63-69, 1987.

[13] J. M. Durante, "Building penetration loss at 900 MHz," *in IEEE Vehicular Technology Conference VTC'73*, pp. 1-7.

[14] P. I. Wells, "The attenuation of UHF radio signals by houses," *IEEE Trans. Vehicular Techn.*, vol. VT-26, no.4, pp. 358-362, Nov. 1977.

[15] E. H. Walker, "Penetration of radio signals into buildings in the cellular radio environment," *Bell Syst. Tech. J.*, vol. 62, no. 9, pp. 2719-2734, Nov. 1983

[16] R. J. Pillmeier, "In-building signal correlation for an urban environment," *in Proc. of 34th IEEE Vehicular Techn. Soc. Conf.*, May 1984, pp. 156-161.

[17] C. Gutzeit and A. Baran, "900 MHz indoor-outdoor propagation investigations via bit error structure measurements," *in Proc. IEEE Vehicular Techn. Conf. VTC'89*, San Francisco, May 1989, pp. 321-328.

[18] S. T. S. Chia and A. Baran, "900 MHz attenuation and bit error ratio measurements inside a modem office building," *in Proc. IEE Fifth Int. Conf. Mobile Radio Personal Commun.*, Warick, U.K., pp. 33-37, Dec. 1989.

[19] D. Molkdar, "Review on radio propagation into and within buildings," *in IEE Proc.- H*, vol. 138, no.1, pp. 61-73, Feb. 1991.

[20] T. S. Rappaport and C. D. McGillem, "Characterizing the UHF factory radio channel," *Electronics Letters*, vol. 23, no.19, pp. 1015-1016, Sept. 10, 1987.

[21] T. S. Rappaport, "Delay spread and time delay jitter for the UHF factory multipath channel," *in Proc. 1988 IEEE Vehicular Technol. Conf. VTC'88*, pp. 186-189, Philadelphia, June 1988.

[22] T. S. Rappaport and C. D. McGillem, "UHF multipath and propagation measurements in manufacturing environments," *in Proc. 1988 IEEE Global Commun. Conference, GLOBECOM' 88*, Hollywood, Fla., Nov. 30, 1988, pp. 825-829.

[23] T. S. Rappaport and C. D. McGillem, "UHF fading in factories," *IEEE J. Selected Areas Communications*, vol. 7, no.1, pp. 40-48, Jan. 1989.

[24] K. Takamizawa, S. Y. Seidel, and T. S. Rappaport, "Indoor radio channel models for manufacturing environments," *IEEE Southeastcon 1989 Proc.*, pp. 750-754, Columbia, S.C., Apr. 10, 1989.

[25] S. Y. Seidel, K. Takamizawa, and T. S. Rappaport, "Application of second-order statistics for an indoor radio channel model," *in IEEE Proc. Vehicular Technology Conf. VTC'89*, San Francisco, pp. 888-892, May 1989.

[26] T. S. Rappaport, "Indoor radio communications for factories of the future," *IEEE Commun. Mag.*, pp. 15-24, May 1989.

[27] T. S. Rappaport, "Characterization of UHF multipath radio channels in factory buildings," *IEEE Trans. Antennas
Propagat.*, vol. 37, vol. 8, pp. 1058-1069, Aug. 1989.

[28] T. S. Rappaport and S. Y. Seidel, "Multipath propagation
models for in-building communications," *in Proc. IEE Fifth
Int. Conf. Mobile Radio and Personal Commun.*, Warick,
pp. 69-74, Dec. 1989.

[29] S. Y. Seidel and T. S. Rappaport, "Simulation of UHF indoor radio channels for open-plan building environments,"
in IEEE Proc. Vehicular Techn. Conf. VTC'90, Miami, Fla.,
pp. 597-602, May 1990.

[30] D. A. Hawbaker and T. S. Rappaport, "Indoor wideband
radio propagation measurement system at 1.3 GHz and 4-0
GHz", *in Proc. IEEE Vehicular Technology Conf. VTC'90*,
Miami, Fla., pp. 626-630, May 1990.

[31] D. A. Hawbaker and T. S. Rappaport, "Indoor wideband
radiowave propagation measurements at 1.3 GHz and 4.0
GHz," *Electronics Letters*, vol. 26, no.1, 1990.

[32] S. Y. Seidel and T. S. Rappaport, "900 MHz path loss measurements and prediction techniques for in-building communication system design," *Proc. 41st Vehicular Techn. Conf.
VTC'91*, Saint Louis, Missouri, May 19-22, 1991, pp. 613-
618.

[33] T. S. Rappaport, S. Y. Seidel, and K. Takamizawa, "Statistical channel impulse response models for factory and open
plan building radio communication system design," *IEEE
Trans. Commun.*, vol. 39, no. 5, pp. 794-807, May 1991.

[34] T. S. Rappaport, "The MPRG research program-some recent
propagation results," *in Virginia Tech's First Symp. Wireless Personal Commun.*, Blacksburg, Va., June 3-5, 1991, pp.
9.1-9.12.

[35] S. Y. Seidel and T. S. Rappaport, "Path loss prediction in multifloored buildings at 914 MHz," *Electron. Letters*, vol. 27, no. 15, July 1991.

[36] T. S. Rappaport and D. A. Hawbaker, "Effects of circular and linear polarized antennas on wideband propagation parameters in indoor radio channels," *in Proc. IEEE GLOBECOM'91 Conf.*, Phoenix, Arizona, Dec. 1991, pp. 1287-1291.

[37] S. Y. Seidel and T. S. Rappaport, "914 MHz path loss prediction models for wireless communications in multifloored buildings," *IEEE Trans. Antennas Propagat.*, vol. 40, no.2, pp. 207-217, Feb. 1992.

[38] S. Y. Seidel, T. S. Rappaport, M. J. Feuerstein, K. L. Blackard, and L. Grindstaff, "The impact of surrounding buildings on propagation for wireless in- building personal communications system design," *in Proc. IEEE Vehicular Techn. Conf., VTC'92*, Denver, Colo., May 1992, pp. 814-818.

[39] T. S. Rappaport, S. Y. Seidel, and K. R. Schaubach, "Site-specific propagation prediction for PCS system design," *in Proc. Second Annual Erginia Tech. Symp. Wireless Personal Commun.*, Blacksburg, Va., June 17-19, 1992.

[40] C. M. P. Ho and T. S. Rappaport, "Effects of antenna polarization and beam pattern on multipath delay spread at 2.45 GHz in indoor propagation channels," *in Proc. 1st Int. Conf. Univ. Personal Commun.*, Dallas, Tex., Sept. 29-Oct. 2, 1992.

[41] T. T. Tran and T. S. Rappaport, "Site specific propagation prediction models for PCS design and installations," *in Milcom'92 Conf.*, San Diego, Calif., Oct. 1992.

[42] S. Y. Seidel and T. S. Rappaport, "A ray tracing technique to predict path loss and delay spread inside buildings," *Proc. IEEE GLOBECOM '92 Conf.*, Orlando, Ha., Dec. 6-9, 1992.

[43] D. M. J. Devasirvatham, "Time delay spread measurements of wideband radio signals within a building," *Electronics Letters*, vol. 20, no. 23, pp. 950-951, Nov. 8, 1984.

[44] D. M. J. Devasirvatham, "Time delay spread measurements of 850 MHz radio waves in building environments," *in Proc. IEEE Global Commun. Conf. GLOBECOM'85*, pp. 970-973, Dec. 1985.

[45] D. M. J. Devasirvatham, "A Comparison of time delay spread measurements within two dissimilar office buildings," *IEEE ICC'86 Proc.*, vol 2, pp. 852-857, Toronto, Canada, June 1986.

[46] D. M. J. Devasirvatham, "Time delay spread and signal level measurements of 850 MHz radio waves in building environments," *IEEE Trans. Antennas Propagat.*, vol. AP-34, no.11, pp. 1300-1305, Nov. 1986.

[47] D. M. J. Devasirvatham, "A comparison of time delay spread and signal level measurements within two dissimilar office buildings," *IEEE Trans. Antenna and Propagat.*, vol. AP-35, no.3, pp. 319-324, Mar. 1987.

[48] D. M. J. Devasirvatham, "Multipath time delay jitter measured at 850 MHz in the portable radio environment," *IEEE J. Select. Areas in Comm.*, vol. SAC-5, no. 5, pp. 855-861, June 1987.

[49] D. M. J. Devasirvatham, "Multipath time delay spread in the digital portable radio environment," *IEEE Commun. Mag.*, vol. 25, no.6, pp. 13-21, June 1987.

[50] D. M. J. Devasirvatham, R. R. Murray, and C. Banerjee, "Time delay spread measurements at 850 MHz and 1.7 GHz inside a metropolitan office building," *Electron. Letters*, vol. 25, no.3, pp. 194-196, Feb. 2, 1989.

[51] D. M. J. Devasirvatham, C. Banerjee, M. J. Krain, and D. A. Rappaport, "Multi-frequency radiowave propagation

measurements in the portable radio environment," *in Proc. IEEE Int. Conf. Commun., ICC'90*, pp. 1334-1340, Apr. 1990.

[52] D. M. J. Devasirvatham, "Multi-frequency propagation measurement and models in a large metropolitan commercial building for personal communications," *in Proc. Second IEEE Int. Symp. Personal, Indoor and Mobile Radio Commun.*, London, England, Sept. 1991, pp. 98-103.

[53] D. M. J. Devasirvatham, C. Banerjee, R. R. Murray and D. A. Rappaport, "Four frequency radiowave propagation measurements of the indoor environment in a large metropolitan commercial building," *in Proc. IEEE GLOBECOM'91 Conf.*, Phoenix, Ariz., Dec. 1991, pp. 1282-1286.

[54] T. A. Sexton and K. Pahlavan, "Channel modeling and adaptive equalization of indoor radio channels," *IEEE J. Selected Areas Commun.*, vol. 7, no.1, pp. 114-120, Jan. 1989.

[55] K. Pahlavan, R. Ganesh, and T. Hotaling, "Multipath propagation measurements on manufacturing floors at 910 MHz," *Electronics Letters*, vol. 25, no.3, pp. 225-227, Feb. 2, 1989.

[56] R. Ganesh and K. Pahlavan, "On amval of paths in fading multipath indoor radio channels," *Electronics Letters*, vol. 25, no.12, pp. 763-765, June 8, 1989.

[57] R. Ganesh and K. Pahlavan, "On the modeling of fading multipath indoor radio channels," *in IEEE GLOBECOM'89 Conference*, Dallas, pp. 1346-1350, Nov. 1989.

[58] S. J. Howard and K. Pahlavan, "Doppler spread measurements of the indoor radio channels," *Electronics Letters*, vol. 26, no.2, pp. 107-109, Jan. 1990.

[59] S. Howard and K. Pahlavan, "Measurement and analysis of the indoor radio channel in the frequency domain," *IEEE Trans. on Instrumentation and Measurement, IM-39*, pp. 75 1-755, Oct. 1990.

[60] K. Pahlavan and S. J. Howard, "Statistical AR models for the frequency selective indoor radio channel," *Electronics Letters*, vol. 26, no.15, pp. 1133-1135, July 1990.

[61] S. J. Howard, K. Pahlavan, "Statistical autoregressive model for the indoor radio channel," *in Proc. IEEE GLOBE-COM'90 Conf.*, San Diego, Calif., Dec. 1990, pp. 1000-1006.

[62] R. Ganesh and K. Pahlavan, "Statistics of short time variations of indoor radio propagation," *in Proc. Int. Conf. Commun. ICC'91*, Denver, Colo., June 23-26, 1991.

[63] R. Ganesh and K. Pahlavan, "Statistical modeling and computer simulation of the indoor radio channel," *in IEE Proc. Part I: Communication, Speech and Vision*, vol. 138, no.3, pp. 153-161, June 1991.

[64] S. J. Howard and K. Pahlavan, "Fading results from narrowband measurements of the indoor radio channel," *in Proc. Second IEEE Int. Symp. Personal, Indoor and Mobile Radio Commun.*, London, England, Sept. 1991, pp. 92-97.

[65] K. Pahlavan and R. Ganesh, "Wideband frequency and time domain models for the indoor radio channel," *in Proc. IEEE GLOBECOM'91 Conference*, Phoenix, Ariz., Dec. 1991, pp. 1135-1140

[66] K. Pahlavan and R. Ganesh, "Statistical characterization of a partitioned indoor radio channel," *in Proc. IEEE Int. Conf. Commun., ICC'92*, Chicago, Ill., June 14-17, 1992, pp. 1252- 1256.

[67] S. J. Howard and K. Pahlavan, "Autoregressive modeling of wide-band indoor radio propagation," *IEEE Trans. Commun.*, vol. 40, no.9, pp. 1540-1552, Sept. 1992.

[68] T. Holt, K. Pahlavan, and J. F. Lee, "Ray tracing algorithm for indoor radio propagation modeling," *in Proceedings Third IEEE Int. Symp. Personal, Indoor and Mobile Radio Commun.*, Boston, Mass., Oct. 19-21, 1992.

[69] R. J. C. Bultitude, "Propagation characteristics of 800/900 MHz radio channels operating within buildings," *in Proc. 13th Biennial Symp. Commun., BI/I -4*, Kingston, Ontario, Canada, June 2-4, 1986.

[70] R. J. C. Bultitude, "Measurement, characterization and modeling of indoor 800/900 MHz radio channels for digital communications," *IEEE Commun. Mag.*, vol. 25, no.6, pp. 5-12, June 1987.

[71] R. J. C. Bultitude and S. A. Mahmoud, "Estimation of indoor 800/900 MHz digital radio channel performance characteristics using results from radio propagation measurements," *in Proc. Int. Commun. Conf. ICC"87*, pp. 70-75, 1987.

[72] R. J. C. Bultitude, S. A. Mahmoud, and W. A. Sullivan, "A comparison of indoor radio propagation characteristics at 910 MHz and 1.75 GHz," *IEEE J. Select. Areas in Comm.*, vol. 7, no.1, pp. 20-30, Jan. 1989.

[73] R.J.C. Bultitude, "Measurements of wideband propagation characteristics for indoor radio with predictions for digital system performance," *in Proc. Wireless '90 Conf.*, Calgary, Alberta, Canada, July 1990.

[74] R. J. C. Bultitude, P. Melancon, R. Hahn, and M. Prokki, "An investigation of static indoor channel multipath characteristics," *in Proc. Wireless '91 Conf.*, July 8-10, 1991, Calgary, Alberta, Canada.

[75] T. Lo, J. Litva, and R. J. C. Bultitude, "High-resolution spectral analysis techniques for estimating the impulse response of indoor radio channels," *in Proc. 1992 IEEE Int. Conf. Selected Topics in Wireless Commun.*, Vancouver, B.C., June 25-26, 1992, pp. 57-60.

[76] H. Hashemi, "Principles of digital indoor radio propagation," *in Proc. IASTED Int. Symp. Computers, Electronics,*

Commun. and Control, CECC'91, Calgary, Alberta, Canada, Apr. 8-10, 1991, pp. 48-53

[77] H. Hashemi, D. Tholl, and G. Morrison, "Statistical Modeling of the indoor radio propagation channel-part I," *in P roc. IEEE Vehicular Techn. Conference, VTC'92,* Denver, Colo., May 1992, pp. 338-342

[78] H. Hashemi, D. Lee, and D. Ehman, "Statistical Modeling of the indoor radio propagation channel-part 11," *in Proc. IEEE Vehicular Technology Conference, VTC '92,* Denver, Colo., May 1992, pp. 839-843.

[79] H. Hashemi, "Impulse response modeling of indoor radio propagation channels," *IEEE J. Selected Areas Commun.,* Sept. 1993, vol. 11, no. 7, pp. 967-978

[80] H. Hashemi and D. Tholl, "Analysis of the rms delay spread of indoor radio propagation channels," *in Proc. IEEE Int. Conf. Commun., ICC '92,* Chicago, Ill., June 14-17, 1992, pp. 875- 881.

[81] D. C. Cox, "Antenna diversity performance in mitigating the effects of portable radiotelephone orientation and multipath propagation," IEEE Trans. Commun., vol. COM-31, pp. 620-628, May 1983.

[82] D. C. Cox, "Time division adaptive retransmission for reducing signal impairments in portable radiotelephones," *IEEE Trans. Vehicular Techn.,* vol. VT-32, 1-10.3, pp. 230-238, Aug. 1983.

[83] D. C. Cox, R. R. Murray, H. W. Amold, A. W. Noms, M. F. Wazowics, "Cross-polarization coupling measured for 800 MHz radio transmission in and around houses and large buildings," *IEEE Trans. Antennas Propagat.,* vol. AP-34, no. 1, pp. 83-87, Jan. 1986.

[84] R. R. Murray, H. W. Amold, and D. C. Cox, "815 MHz radio attenuation measured within a commercial building,"

IEEE Antennas Propagat. Int. Symp., vol. 1, pp. 209-212, Philadelphia, June 8-13, 1986.

[85] A. A. M. Saleh and R. A. Valenzuela, "A statistical model for indoor multipath propagation," *in Proc. Int. Conf. Commun., ICC'86*, pp. 837-841, Toronto, June 1986.

[86] A. A. M. Saleh and R. A. Valenzuela, "A statistical model for indoor multipath propagation," *IEEE J. Selected Areas in Comm.*, vol. SAC-5, no.2, pp. 128-137, Feb. 1987.

[87] A. A. M. Saleh, A. J. Rustako, Jr., and R. S. Roman, "Distributed antennas for indoor radio communications," *in P roc. Int. ConCommun., ICC'87*, pp. 76 – 80, 1987.

[88] A. A. M. Saleh, A. J. Rustako, Jr., and R. S. Roman, "Distributed antennas for indoor radio communications," *IEEE Trans. Commun.*, vol. COM- 35, 110.12, pp. 1245-1251, Dec. 1987.

[89] A. J. Motley and D. A. Palmer, "Directed radio coverage within buildings," *in IEE Conf. Publication*, no.224, pp. 56-60, Sept. 1983.

[90] A. J. Motley and D. A. Palmer, "Reduced long-range signal reception with leaky feeders," *Electronics Letters*, vol. 19, no.18, pp. 714-715, Sept. 1983.

[91] D. A. Palmer and A. J. Motley, "Controlled radio coverage within buildings," *British Telecomm. Technol. J.*, vol. 4, no.4, pp. 55-57, Oct 1986

[92] A. J. Motley and J. M. P. Keenan, "Personal communication radio coverage in buildings at 900 MHz and 1700 MHz," *Electron. Letters*, vol. 24, no.12, pp. 763-764, June 9, 1988.

[93] A. J. Motley and A. J. Martin, "Radio coverage in buildings," *in Proc. Nut. Commun. Forum, NCF'88*, Chicago, 1988, pp. 1722-1730

[94] S. E. Alexander, "Radio propagation within buildings at 900 MHz," *Electron. Letters*, vol. 18, no.21, pp. 913-914, Oct. 14, 1982.

[95] S. E. Alexander and G. Pugliese, "Cordless communication within buildings: Results of measurements at 900 MHz and 60 GHz," *British Telecom. Technol. J.*, vol. I , no. I , pp. 99-105, July 1983.

[96] S. E. Alexander, "Characterizing buildings for propagation at 900 MHz," *Electron. Letters*, vol. 19, p. 860, Sept. 29, 1983.

[97] S. E. Alexander, "Radio propagation within buildings at 900 MHz," *in IEE Third Int. Conf. Antennas Propagat.*, 1983, pp. 177-180

[98] S. E. Alexander, "900 MHz propagation within buildings," *in IEE Second Conf. Radio Spectrum Conservation Techn.*, May 1984, pp. 51-55

[99] P. J. Barry and A. G. Williamson, "UHF radiowave penetration into buildings and signal level variations within," *in Proc. 23rd New Zealand Nar. Electronics Conf.*, Palmerston North, New Zealand, Aug. 26-28, 1986, pp. 94-100.

[100] P. J. Barry and A. G. Williamson, "UHF radiowave penetration into buildings and propagation within," *in VHF and UHF Radio Syst. Symp.*, Nov. 27-28, 1988, pp. 32-38.

[101] P. J. Barry and A. G. Williamson, "Statistical model for UHF radio-wave signals within extremally illuminated multistory buildings," emphin IEE Proceedings-I, vol. 138, no.4, 1991.

[102] H. Zaghloul, G. Morrison, D. Tholl, R. J. Davies, S. Kazeminejad, "Frequency response measurements of the indoor channel," *in Proc. ANTEM'90 Conf*, Winnipeg, Manitoba, Aug. 1990, pp. 267-272

[103] M. Fattouche, G. Morrison, H. Zaghloul, L. Petherick, "Diversity for indoor radio communications," *Proc. 33rd Midwest Symp. Circuits Syst.*, Aug. 12-14, 1990, Calgary, Alberta, Canada.

[104] H. Zaghloul, G. Morrison, M. Fry, and M. Fattouche, "Measurements of the frequency response of the indoor channel," emphin Proc. 33rd IEEE Midwest Symp. Circuits Syst., Aug. 12- 14, 1990.

[105] G. Momson, H. Zaghloul, M. Fattouche, M. Smith, and A. McGirr, "Frequency measurements of the indoor channel: System evaluation and post processing using IDFT and ARMA modeling," *in Proc. of the IEEE Pacific Rim Conf Commun., Computers and Signal Processing*, May 9- 10, 1991.

[106] H. Zaghloul, G. Morrison, and M. Fattouche, "Frequency response and path loss measurements of indoor channel," *Electron. Letters*, vol. 27, no.12. June 1991.

[107] G. Morrison, M. Fattouche, H. Zaghloul, and D. Tholl, "Frequency measurements of the indoor channel," *in Proc. Wireless'91 Conf.*, Calgary, Alberta, Canada, July 8-10, 1991.

[108] G. Momson, M. Fattouche, and D. Tholl, "Parametric modeling and spectral estimation of indoor radio propagation data," *in Proc. 'Wireless'92 Conf.*, Calgary, Alberta, Canada, July 1992.

[109] G. Morrison, M. Fattouche, and H. Zaghloul, "Statistical analysis and autoregressive modeling of the indoor channel," *in Proc. 1st Int. Conf. Universal Personal Commun.*, Dallas, Tex., Sept. 29 - Oct 2, 1992.

[110] P. W. Huish, "Personal communications systems-new requirements for antenna and propagation knowledge," *in 6th Int. Conf. on Antennas and Prop., ICAP'89, Part 2: Propagation*, pp. 371-376

[111] P. W. Huish and G. Pugliese, "60 GHz radio system for propagation studies in buildings," *in IEE Third Int. Conf Antennas Propagat..* 1983. pp. 181-185.

[112] G. Pugliese, "A 60 GHz radio system for propagation studies in buildings," *in IEE Third Int. Conf. Antennas Propagat.*, 1983.

[113] M. Kaji, "UHF-band radio propagation characteristics within large buildings," *Trans. Inst. Electron. Inf. Commun. Eng.*, vol. J 70-B, no.10, pp. 1200-1209, Oct. 1987.

[114] M. Kaji, "Polarization characteristics in UHF-band mobile radio propagation," *Trans. Inst. Electron. Inf. Commun. Eng.*, vol. J 70-B, no.12, pp. 1510-1521, Dec. 1987.

[115] M. Kaji, "UHF-band path loss prediction within small buildings by ray method," *Trans. Inst. Electron. Inf. Commun. Eng.*, vol. J 71-B, no.1, pp. 89-91, Jan. 1988.

[116] A. M. D. Turkmani and A. F. Toledo, "Radio transmission at 1800 MHz into and within multistory buildings," *in IEE Proc.-Part I*, vol. 138, no.6, pp, 577-584, Dec. 1991.

[117] Y. Yamaguchi, T. Abe, and T. Sekiguchi, "Experimental study of radio propagation characteristics in an underground street and condors," *IEEE Trans. Electromagn. Compat.*, vol. EMC-28, no.3, pp. 148-155, Aug. 1986.

[118] Y. Yamaguchi, T. Abe, and T. Sekiguchi, "Radio propagation characteristics in underground streets crowded with pedestrians," *IEEE Tranv. Electromagn. Compat.*, vol. 30, no.2, pp. 130-136, May 1988.

[119] P. F. M. Smulders and A. G. Wagemans, "Wideband indoor radio propagation measurement at 58 GHz," *Electron. Letters*, vol. 28, no. 13, 1992.

[120] P. F. M. Smulders and A. G. Wagemans, "Wideband measurements of mm-wave indoor radio channels," *in Proc.*

Third IEEE Int. Symp. Personal, Indoor and Mobile Radio Commun., Boston, Mass., Oct. 19-21, 1992.

[121] P. F. M. Smulders and A. G. Wagemans, "A statistical model for the mm-wave indoor radio channel," *in Proc. Third IEEE Int. Symp. Personal, Indoor and Mobile Radio Commun.*, Boston, Mass., Oct. 19-21, 1992.

[122] S. R. Todd, M. El-Tanany, and S. A. Mahmoud, "Space and frequency diversity measurements of the 1.75 GHz indoor radio channel for portable data terminals and telephones," *in Proc. IEEE Vehicular Techn. Conf., VTC'92*, Denver. Colo., May 1992 pp. 613-616.

[123] G. A. Kalivas, M. El-Tanany, and S. A. Mahmoud, "Millimeter- communications," *in Proc. IEEE Vehicular Technology Conf. VTC'92*, Denver, Colo., May 1992, pp. 609-612.

[124] S. R. Todd, M. S. El-Tanany, and S. A. Mahmoud, "Four branch diversity measurements at 1.7 GHz for indoor wireless communications," *in Proc. 1992 IEEE Int. Conf. Selected Topics in Wireless Commun.*, Vancouver, B.C., June 25-26, 1992, pp. 69-72.

[125] S. R. Todd, M. S. El-Tanany, and S. A. Mahmoud, "Space and frequency division measurements of the I .7 GHz indoor radio channel using a four-branch receiver," *in IEEE Trans. Vehic. Techn.*, vol. 41, no.3, pp. 312-320, Aug. 1992.

[126] S. R. Todd, M. S. El-Tanany and S. A. Mahmoud, "Space and frequency diversity measurements of the 1.7GHz indoor radio channel for wireless personal communications," *in Proc. 1st Int. Conf. Universal Personal Commun.*, Dallas, Tex., Sept. 29-Oct. 2, 1992.

[127] P. Karlsson, "Investigation of radio propagation and macroscopic diversity in indoor microcells at 1700 MHz," *in Proc. IEEE Vehicular Techn. Conf., VTC'90*, pp. 390-395, Miami, May 1990.

[128] P. Karlsson and H. Borjesson, "Measurement system for in-
      door narrowband radio propagation at 1700 MHz and some
      results," *in Proc. IEEE Vehicular Techn. Conf. VTC'92*,
      Denver, Colo., May 1992, pp. 625-628.

[129] P. F. Driessen, M. Gimersky, and T. Rhodes, "Ray model of
      indoor propagation," *in Proc. Second Annual Virginia Tech.
      Symp. Wireless Personal Commun.*, Blacksburg, Va., June
      17- 19, 1992.

[130] P. F. Driessen, "Development of a propagation model in
      the 20-60 GHz band for wireless indoor communications,"
      *in Proc. IEEE Pacific Rim Conf. Communications, Com-
      puters, and Signal Processing*, Vancouver, British Columbia,
      Canada, May 1991, pp. 59-62.

[131] P. Melancon, "Report on propagation inside an empty and
      furnished building at 433, 861, and 1705 MHz," *Queens Uni-
      versity 15th Biennial Symp. Communications*, Queens Univ.,
      Kingston, Ontario, Canada, June 3-6, 1990, pp. 81-84.

[132] J. LeBell and P. Melancon, "The development of a compre-
      hensive indoor propagation model," *in Proc. Second IEEE
      Int. Symp. Personal, Indoor and Mobile Radio Communica-
      tions*, London, England, Sept. 1991, pp. 75-79.

[133] J. LeBell and P. Melancon, "The development of a compre-
      hensive indoor propagation model," *in Proc. Second IEEE
      Int. Symp. Personal, Indoor and Mobile Radio Communica-
      tions*, London, England, Sept. 1991, pp. 75-79.

[134] L. W. Pickering, E. H. Bamhart, M. L. Witten, N. H.
      Hightower, and M. D. Frerking, "Characterization of indoor
      propagation for personal communication services," *in IEEE
      Southcon'91 Conf. Record*, Mar. 1991.

[135] L. W. Pickering, E. N. Bamhart, and M. L. Witten, "Sta-
      tistical data from frequency domain measurements of indoor
      PCN communication channel," *in Proc. Second IEEE Int.*

Symp. Personal, Indoor and Mobile Radio Communications, London, England, Sept. 1991, pp. 86-91.

[136] L. W. Pickeiry, E. N. Bamhart, and M. L. Witten, "Measurements of the multipath spread of the indoor wireless communication channel," *in Proc. Third IEEE Int. Symp. Personal, Indoor and Mobile Radio Commun.*, Boston, Mass., Oct. 19-21, 1992.

[137] A. Angus, C. Tannous, and B. Davies, "Chaos in radio communication," *in Proc. Wireless'90 Conf.*, Calgary, Alberta, July 1990.

[138] C. Tannous, R. Davies, and A. Angus, "Strange attractors in multipath propagation," *IEEE Trans. Commun.*, vol. 39, no.5, pp. 629-631, May 1991.

[139] S. A. Bergmann and H. W. Arnold, "Polarization diversity in portable communications environment," *Electron. Letters*, vol. 22, no. 1 I , pp. 609-610, May 22, 1986.

[140] P. Yegani, "On the probability density of the power spectrum of the transfer function for a two-path and three-path factory radio channel," *in Proc. 41st Vehicular Techn. Conf., VTC '91*, Saint Louis, Missouri, May 19-22. 1991, pp. 477-481.

[141] General Electric System Application Manual, Sec. 80-AI, Table IV-3. Lynchburg, VA: General Electric Corporation, Dec. 1972.

[142] J. Shefer, "Propagation statistics of 900 MHz and 450 MHz signals inside buildings," *in Microwave Mobile Radio Symp.*, Boulder, Colo., Mar. 7-9, 1973.

[143] H. Kishimoto, "Indoor radio propagation analysis by ray method," *IECE of Japan, Tech. Rep. on Antennas Propagat.*, A.P76-62, 1976.

[144] M. Komura, T. Hogihira, and M. Ogasawara, "New radio paging system and its propagation characteristic," *IEEE*

Trans. Vehicular Techn., vol. VT-26, no. 4, pp. 362-366, Nov. 1977.

[145] K. Tsujimura and M. Kuwabara, "Cordless telephone system and its propagation characteristics," *IEEE Trans. Vehicular Techn.*, vol. VT-26, no. 4, pp. 367-371, Nov. 1977.

[146] J. H. Winters and Y. S. Yeh, "On the performance of wideband digital radio transmission within buildings using diversity," *in Proc. GLOBECOME '85 Conf.* New Orleans, Louisiana, Dec. 1985 pp. 991-996

[147] J. Horikoshi, K. Tanaka, T. Morinaga, "1.2 GHz band wave propagation measurements in concrete buildings for indoor radio communications," *IEEE Trans. Vehicular Techn.*, vol. VT- 35, no. 4, pp. 146-152, Nov. 1986.

[148] P. Yegani and C. D. McGillem, "A statistical model for line-of sight (LOS) factory radio channels," *in Proc. Vehicular Techn. Conf. VTC'89*, pp. 496-503, San Francisco, May 1989.

[149] P. L. Camwell and J. C. McRory, "Experimental results of in-building anisotropic propagation at 835 MHz using leaky feeders and dipole antennas," *in Proc. MONTECH '87 Conf. Commun.*, pp. 213-216, 1987.

[150] K. J. Bye, "Leaky-feeders for cordless communication in the office," *in Proc. EUROCON'88 Conf.*, 1988, pp. 387-390.

[151] A. R. Tharek and J. P. McGeehan, "Indoor propagation and bit error rate measurements at 60 GHz using phase-locked oscillators," *in Proc. IEEE Vehicular Techn. Conf., VTC '88*, Philadelphia, June 1988, pp. 127-133.

[152] D. Akerberg, "Properties of a TDMA pico cellular office communication system." *in IEEE GLOBECOM '88*. Hollywood. Fla., Dec. 1988, pp. 1343-1349.

[153] C. Bergljung, et al., "Micro-cell radio channel. Preliminary report on indoor field-strength measurements at 900 and

1700 MHz," Dept. of Applied Electronics, Lund University, Lund, Jan. 19, 1989.

[154] P. Yegani and C. D. McGillem, "A statistical model for the obstructed factory radio channel," *in Proc. Global Commun. Conf. GLOBECOM'89.* pp. 1351-1355. Dallas. Nov. 1989.

[155] T. K. Ishii, "RF propagi6on in buildings," emph Radio Frequency Design, vol. 12, Part 7, pp. 45-49, July 1989.

[156] F. C. Owen and C. D. Pundey, "Radio propagation for digital cordless telephones at 1700 MHz and 900 MHz," *Electron. Letters*, vol. 25, no. I, pp. 52- 53, Jan. 5, 1989.

[157] F. C. Owen and C. D. Pundey, "In-building propagation at 900 MHz and 1650 MHz for digital cordless telephone," *in Sixth Int. Conf. Antenna Propagat., ICAP'89*, Part 2: Propagation, 1989, pp. 276-281.

[158] R. J. Bailey and G. R. Summers, "Radio channel characterization for the digital European cordless telecommunications system," *British Telecom. J.*, vol. 8, no. I, pp. 25-30, Jan. 1990.

[159] J. F. Lafortune, and M. Lecours, "Measurement and modeling of propagation losses in a building at 900MHz," *IEEE Trans. Vehicular Techn.*, vol. 39, no. 2, pp. 101-108, May 1990.

[160] P. Yegani and C. D. McGillem, "A statistical model for the factory radio channel," *IEEE Trans. on Communications*, vol. 39, no. 10, pp. 1445-1454, Oct. 1991.

[161] T. Takeuchi, M. Sako, and S. Yoshida, "Multipath delay estimation for indoor wireless communication," *in Proc. IEEE Vehicular Techn. Conf., VTC'90*, pp. 401-406, 1990.

[162] M. R. Heath, "Propagation measurements at 1.76 GHz for digital European cordless telecommunications," *in Proc. IEEE GLOBECOM '90 Conf.*, San Diego, Calif., Dec. 1990, pp. 1007-1012

[163] R. Davies, M. Bensebti, M. Beach, J. P. McGeehan, D. Rickard, C. Shepherd, and S. Wales, "A comparison of indoor and urban propagation at 1.7, 39, and 60 GHz," *in Proc. of the 41st Vehicular Technology Conference, VTC'91,* Saint Louis, Missouri, May 19-22, 1991, pp. 589-593.

[164] J. R. Barry, J. M. Kahn, E. A. Lee, and D. G. Messerschmitt, "Simulation of multipath impulse response for indoor diffuse optical channels," *in Proc. IEEE Workshop on Wireless LAN's,* Worcester, MA, May 9-10, 1991, pp. 81- 87.

[165] G. Heidari and C. D. McGillem, "Performance limitations of the indoor radio channel," *in Proc. Second IEEE Int. Symp. Personal, Indoor and Mobile Radio Commun.,* London, England, Sept. 1991, pp. 80-85.

[166] M. C. Lawton, R. L. Davies, and J. P. McGeehan, "A ray launching method for the prediction of indoor radio channel characteristics," *in Proc. Second IEEE Int. Symp. Personal, Indoor and Mobile Radio Commun.,* London, England, Sept. 1991, pp. 104-108.

[167] J. W. McKown and R. L. Hamilton, "Ray tracing as a design tool for radio networks," *IEEE Network Mag.,* vol. 5, no. 6, pp. 27-30, Nov. 1991

[168] J. E. Mitzlaff, "Radio propagation and anti-multipath techniques in the WIN environment," *IEEE Network Mag.,* vol. 5, no. 6, pp. 21-26, Nov. 1991

[169] R. A. Ziegler and J. M. Cioffi, "Estimation of time-varying digital mobile radio channel," *in Proc. IEEE GLOBECOM '91 Conf.,* Phoenix, Ariz., Dec. 1991, pp. 1130-1134.

[170] G. G. M. Janssen and R. Prasad, "Propagation measurements in indoor radio environments at 2.4 GHz, 4.75 GHz and 11.5 GHz," *in Proc. IEEE Vehicular Techn. Conf., VTC '92,* Denver, Colorado, May 1992, pp. 617-620.

[171] F. Lotse, J-E. Berg, and R. Bownds, "Indoor propagation measurement at 900 MHz," *in Proc. IEEE Vehicular Techn. Conf., VTC '92*, Denver, Colo., May 1992, pp. 629-632.

[172] S. Harbin, C. Palmer, and B. K. Rainer, "Measured propagation characteristics of simulcast signals in an indoor microcellular environment," *in Proc. IEEE Vehicular Techn. Conf, VTC '92*, Denver, Colorado, May 1992, pp. 604-608.

[173] E. Moriyama, M. Mizuno, Y. Nagata, Y. Furuya, I. Kamiya, and S. Hattori, "2.6 GHz land multipath characteristics measurement in a shielded building," *in Proc. IEEE Vehicular Techn. Conf., VTC '92*, Denver, Colo., May 1992, pp. 621-624.

[174] C. C. Huang and R. Khayata, "Delay spreads and channel dynamics measurements at ISM bands," *in Proc. IEEE Int. Conf. Commun., ICC '92*, Chicago, Ill., June 14-17, 1992, pp. 1222-1226.

[175] W. A. McGladdery and S. Stapleton, "Investigation of polarization effects in indoor radio propagation," *in Proc. 1992 IEEE Int. Conf. Selected Topics in Wireless Commun.*, Vancouver, B. C., June 25-26, 1992, pp. 53-56.

[176] D. I. Laurenson, A. U. H. Sheikh, and S. McLaughlin, "Characterization of the indoor mobile channel using a ray tracing technique," *in Proc. 1992 IEEE Int. Conf. Selected Topics in Wireless Commun.*, Vancouver, B.C., June 25-26, 1992, pp. 65-68.

[177] R. H. S. Hardy and E. Lo, "Propagation coverage prediction techniques for indoor wireless communications," *in Proc. 1992 IEEE International Conf. Selected Topics in Wireless Commun.*, Vancouver, B.C., June 25-26, 1992, pp. 73-75.

[178] W. Honcharenko, H. L. Bertoni, and J. Dailing, "Theoretical prediction of UHF propagation within office buildings," *in Proc. 1st Int. Conf. Universal Personal Commun.*, Dallas, Tex., Sept. 29-Oct. 2, 1992.

[179] J. F. Kiang, "Geometrical ray tracing approach for indoor wave propagation in a condor," *in Proc. 1st Int. Conf. Universal Personal Commun.*, Dallas, Texas, Sept. 29-Oct. 2, 1992.

[180] G. Vannucci and R. S. Roman, "Measurement results on indoor radio frequency re-use at 900 MHz and 18 GHz," *in Proc. Third IEEE Int. Symp. Personal, Indoor and Mobile Radio Commun.*, Boston, Mass., Oct. 19-21, 1992.

[181] R. Khayata and C. C. Huang, "Characterizing wireless indoor communications: measurements in the ISM bands with a directional antenna," *in Proc. Third IEEE Int. Symp. Personal, Indoor and Mobile Radio Commun.*, Boston, Mass., Oct. 19-21, 1992.

[182] L. Van Der Jagt, G. J. Martin, M. A. Maslied, O. L. Storoshchuk, and B. Szabados, "Propagation measurements at a G.M. manufacturing plant for wireless LAN communication," *in Proc. Third IEEE International Symposium on Personal, Indoor and Mobile Radio Commun.*, Boston, Mass., Oct. 19-21, 1992.

[183] Khoi D. Le, Michael W. Hoffman, and Robert D. Palmer, "A stochastic image-based indoor channel model for use in ultra-wideband 3-D sensor array simulations", *2003 IEEE Conference on Ultra Wideband Systems and Technologies*, 2003, pp. 300-304.

[184] C. C. Chong, C. M. Tan, D. I. Laurenson, S. McLaughlin, M. A. Beach and A. R. Nix"A Novel Wideband Dynamic Directional Indoor Channel Model Based on a Markov Process", *IEEE Transactions on Wireless Communications*, vol. 4, no. 4, July 2005, pp. 1539-1552.

[185] Hongbing Zhang and H. H. Fan,"An Indoor Wireless channel Model Based on Wavelet Packets", *Thirty-Fourth Asilomar Conference on Signals, Systems and Computers*, 29 Oct 2000-01 Nov 2000, pp. 455-459.

[186] G. D. Kondylis, F. De Flaviis, G. J. Pottie, and Y. Rahmat-Samii, "Indoor Channel Characterization for Wireless Communications Using Reduced Finite Difference Time Domain (R-FDTD)", *Vehicular Technology Conference, 1999. VTC 1999 - Fall*, pp. 1402-1406.

[187] J. H. Tarng, Wen-Shun Liu, Yeh-Fong Huang, and Jiunn-Ming Huang, "A Novel and Efficient Hybrid Model of Radio Multipath-Fading Channels in Indoor Environments", *IEEE Trans. on Ant. and Propagat.*, vol. 51, no. 3, March 2003, pp. 585-594

[188] Chia-Chin Chong, Youngeil Kim, and Seong-Soo Lee, "A Modified S-V Clustering Channel Model for the UWB Indoor Residential Environment", *IEEE Vehicular Technology Conference, 2005. VTC 2005-Spring*, 2005, pp. 58-62.

[189] R. Tingley and K. Pahlavan, "A statistical model of space-time radio propagation in indoor environments", *IEEE-APS Conference on Antennas and Propagation for Wireless Communications*, 2000, pp. 61-64.

[190] P. Sonthikorn and O. K. Tonguz, "A New Multiple Scatterer Model for Fixed Indoor Wireless Communication Channels", *IEEE International Conference on Communications, 2007. ICC'07*, 2007, pp. 5028-5033.

[191] M. Alsehaili, S. Noghanian, D. A. Buchanan, and A. R. Sebak, "Angle-of-Arrival Statistics of a Three-Dimensional Geometrical Scattering Channel Model for Indoor and Outdoor Propagation Environments", *IEEE Antennas and Wireless Propagation Letters*, 2010, pp. 946-949.

[192] S. Wang, K. Raghukumary, A. Abdi, J. Wallacez, and M. Jensenz, "Indoor MIMO Channels: A Parametric Correlation Model and Experimental Results", *IEEE Symp. on Advances in Wired and Wireless Communication*, 2004, pp. 1-5.

[193] Z. Irahhauten and H. Nikookar, "On the Frequency Dependence of Wireless Propagation Channel's Statistical

Characteristics", *IEEE Vehicular Technology Conference, 2003. VTC 2003-Spring*, 22-25 April 2003, pp. 222-226.

[194] A. Vigil, "Real-time emulation of indoor multipath propagation channels using statistical and deterministic FIR models and acoustic charge transport technology", *IEEE Vehicular Technology Conference, 1993*, 1993, pp. 223-226.

[195] D. Cassioli, M. Z. Win and A. F. Molisch, "A Statistical Model for the UWB Indoor Channel", *IEEE Vehicular Technology Conference, 2001*, 06 May 2001-09 May 2001, pp. 1159-1163.   .

[196] X. Zhao, J. Kivinen and P. Vainikainen, "Tapped Delay Line Channel Models at 5.3 GHz in Indoor Environments", *IEEE Vehicular Technology Conference, 2000*, 24 Sep 2000-28 Sep 2000, pp. 1-5.

[197] D. Cassioli, M. Z. Win and A. F. Molisch, "The Ultra-Wide Bandwidth Indoor Channel: From Statistical Model to Simulations", *IEEE Journal on Selected Areas in Communications*, vol. 20, no. 6, Aug 2002, pp. 1247-1257.

[198] A. F. Molisch, J. R. Foerster and M. Pendergrass, "Channel models for ultra-wideband personal area networks", *IEEE Wireless Commun. Mag.*, vol. 10, no. 6, Dec 2003, pp. 14-21.

[199] S. S. Ghassemzadeh, L. J. Greenstein, T. Sveinsson, A. Kavcic, and V. Tarokh, "UWB indoor delay profile model for residential and commercial environments", *IEEE Veh. Technol. Conf.-Fall*, Oct. 2003, vol. 5, pp. 3120-3125.

[200] H. Nikookar and H. Hashemi, "Statistical modeling of signal amplitude fading of indoor radio propagation channel", *IEEE Inter. Conf. on Unives. Person. Commun.*, vol. 1, pp. 84-88, Oct. 1993.

[201] B. Sujak, D. K. Ghodgaonkar, B. M. Ali and S. Khatun,"Indoor Propagation Channel Models for WLAN

802.1lb at 2.4GHz ISM Band", *IEEE Asia-Pacific Conference on Applied Electromagnetics, 2005. APACE 2005*, 20-21 Dec. 2005.

[202] D. Mavrakis and S. R. Saunders, "A delay-centred wideband indoor channel model for MM-wave communications", *IEEE International Conference on Antennas and Propagation*, 31 March-3 April 2003, pp. 518-521.

[203] A. F. Molisch, J. R. Foerster and M. Pendergrass, "Channel models for ultra-wideband personal area networks", *IEEE Wireless Commun. Mag.*, vol. 10, no. 6, Dec 2003, pp. 14-21.

[204] A. F. Molisch, J. R. Foerster and M. Pendergrass, "Channel models for ultra-wideband personal area networks", *IEEE Wireless Commun. Mag.*, vol. 10, no. 6, Dec 2003, pp. 14-21.

[205] M. Fattouche, L. Petherick, and A. Fapojuwo, "Diversity for mobile radio communications", *Proc. 15th Biennial Symp. Commun.*, Kingston, Ontario, June 1990, pp. 196-199.

[206] H. Hashemi, "Simulation of the urban radio propagation channel", *IEEE Trans. Vehicular Techn.*, vol. VT-28, pp. 213-224, Aug. 1979.

[207] H. Nikookar and H. Hashemi, "Phase Modeling of Indoor Radio Propagation Channels", *IEEE Trans. Vehicular Techn.*, vol. 49, no. 2, March 2000, pp. 594-606.

[208] A. M. Magableh and M. M. Matalgah, "Channel Characteristics of the Generalized Alpha-Mu Multipath Fading Model", *IEEE Wireless Communications and Mobile Computing Conference (IWCMC), 2011*, 4-8 July 2011, pp. 1535 - 1538.

[209] K. Peppas, F. Lazarakis, A. Alexandridis, and K. Dangakis, "Error Performance of Digital Modulation Schemes with MRC Diversity Reception over η-μ Fading Channels",

IEEE Trans. on Wireless Comms., vol. 8, no. 10, Oct. 2009, pp. 4974-4980.

[210] N. Y. Ermolova, "Moment Generating Functions of the Generalized $\eta - \mu$ and $\kappa - \mu$ Distributions and Their Applications to Performance Evaluations of Communication Systems", *IEEE Comms. Letters*, vol. 12, no. 7, Jul. 2008, pp. 502-504.

[211] Saad Al-Ahmadi, and Halim Yanikomeroglu, "On the Approximation of the Generalized-K Distribution by a Gamma Distribution for Modeling Composite Fading Channels", *IEEE Trans. on Wireless Comms.*, vol. 9, no. 2, Feb. 2010, pp. 706-713.

[212] P. Beckmann and A. Spizzichino, "The Scattering of Electromagnetic Waves from Rough Surfaces", Pergamon, New York, 1963.

[213] H. Suzuki, "A statistical model for urban radio propagation", *IEEE Trans. Commun.*, vol. COM-25, pp. 673-680, July 1977.

[214] R. W. Lorenz, "Theoretical distribution functions of multipath propagation and their parameters for mobile radio communication in quasi- smooth terrain", *NATO AGARD Conference Publication 269-17*, Sept. 1979.

[215] N. H. Shepherd, "Radio wave loss deviation and shadow loss at 900 MHz", *IEEE Trans. Vehicular Techn.*, vol. 26, no. 4, pp.309-313, Nov. 1977.

[216] S. O. Rice, "Mathematical analysis of random noise", *Bell Syst. Tech. J.* , vol. 23, pp. 282-332, 1944, and vol. 24, pp. 46-156, 1954.

[217] E. N. Gilbert, "Capacity of a burst-noise channel," *Bell System Tech. J.*, pp. 1253-1265, 1960.

[218] Min-Te Chao, "Statistical properties of Gilbert's burst noise model," *Bell System Tech. J.*, pp. 1303-1324, Oct. 1973.

[219] A. Ekholm, "A generalization of the two-state two-interval semi-Markov model," *in Stochastic Point Processes*, edited by P. A. W. Lewis. New York: Wiley, 1972, pp. 272-284

[220] A. Ekholm, "A pseudo-Markov model for stationary series of events," *Commentations Physico-Mathematicue*, vol. 41, pp. 73-121, 1971.

[221] G. L. Turin, et al., "A statistical model of urban multipath propagation," *IEEE Trans. Vehicular Techn.*, vol. VT-21, pp. 1-9, Feb. 1972

[222] Luoquan Hu and Hongbo Zhu, "Bounded Brownian bridge model for UWB indoor multipath channel," *IEEE International Symposium on Microwave, Antenna, Propagation and EMC Technologies for Wireless Communications, 2005. MAPE 2005*, 8-12 Aug. 2005, pp. 1411-1414.

[223] N. Azzaoui, L. Clavier, and R. Sabre, "Path delay model based on α-stable distribution for the 60GHz indoor channel," *IEEE Global Telecommunications Conference, 2003. GLOBECOM*, 1-5 Dec. 2003, pp. 1638-1643.

[224] A. Knopp, C. A. Hofmann et al., "Extension of Indoor SISO Propagation Models for Correlated MIMO Channels-An Exemplification Applying Saleh's Model", *IEEE Intern. Symp. on Personal, Indoor and Mobile Radio Comm.*, pp. 1-6, 2007.

[225] J. Kermoal, L. Schumacher, K. Pedersen, P. Mogensen, and F. Frederiksen, "A stochastic MIMO radio channel model with experimental validation,", *IEEE Journal on Selected Areas in Comm.*, vol. 20, no. 6, pp. 1211-1226, 2002.

[226] H. Bolcskei et al., "Space-time wireless systems: from array processing to MIMO communications", Cambridge University Press, 2006.

[227] W. Weichselberger et al., "A stochastic MIMO channel model with joint correlation of both link ends", *IEEE Trans. on Wireless Comms.*, vol. 5, no. 1, pp. 90-100, 2006.

[228] A. Saleh and R. Valenzuela, "Statistical Model for Indoor Multipath Propagation,", *IEEE Journal on Selected Areas in Comm.*, vol. 5, no. 2, pp. 128-137, 1987.

[229] J. Wallace and M. Jensen, "Modeling the indoor MIMO wireless channel,", *IEEE Trans. on Antennas and Propagation*, vol. 50, no. 5, pp. 591-599, 2002.

[230] T. Zwick, C. Fischer, and W. Wiesbeck, "A stochastic multipath channel model including path directions for indoor environments", *IEEE Journal on Selected Areas in Comm.*, vol. 20, no. 6, pp. 1178-1192, 2002.

3

Acoustic Wave Propagation

CONTENTS

DOI: 10.1201/9781003213017-3

Acoustic waves are the choicest mode of communication in underwater and underground scenarios. Acoustic waves are longitudinal waves that travel through the medium by adiabatic compression and relaxation of the collection of particles in the medium. Acoustic waves are characterized by acoustic pressure, particle velocity, particle displacement and acoustic intensity. Rays can be used to represent acoustic waves emitted from a source while propagating in different directions. Sums of the contributions from these rays is used to calculate the acoustic pressure field at the receiver. the acoustic pressure field existing between the source and the receiver are referred to as eigenrays. The eigenrays follow straight lines if the medium is homogeneous. The rays are refracted on the way if the medium is heterogeneous and the velocity of the acoustic waves varies with distance. The flow of acoustic waves is guided by Eikonal equation. The Eikonal equation is a first-order non-linear partial differential equations (PDE) that can be solved by different techniques. The most convenient of these is to use the method of characteristics. In this case, a family of curves (rays) are introduced, which are perpendicular to the level curves (wave-fronts). This family of rays defines a new coordinate system, and it turns out that in ray coordinates the Eikonal equation reduces to linear ordinary differential equation. However, modeling the interaction within the underwater or underground environment is extremely challenging. Therefore, different techniques are used in different scenarios. This chapter elaborates on different techniques that are used for modeling acoustic wave propagation and the impact of the environment on their movement.

3.1   Introduction

If we want to communicate in scenarios like underwater or underground, we need a mode or carrier of information that can propagate through this unconventional media. Underwater or underground channels are quite different from classical wireless channels as seawater is conductive and solid ground is absorptive resulting in exponential attenuation of electromagnetic field strength. Another problem is the skin effect when the EM waves tend to concentrate only in the upper layer or skin depth of water or soil instead of the central core path depending on the frequency of operation. Skin depth is given by $\delta = \sqrt{2/2\pi f \mu \sigma}$, where σ is the electrical conduction, f is the frequency of operation, and μ is the magnetic permeability. As a result, EM waves can only travel over a range of a few 10's of meters and the range depends on the frequency of operation. Higher the frequency, lower is the range of signal propagation. It's only acoustic pressure waves that are capable of travelling through this non-traditional media experiencing relatively low attenuation [1].

There is no typical underwater or underground communication environment as different geographical areas are affected by different physical parameters and processes. Therefore system developers resort to extensive site-specific measurement campaigns, trials, testings, and validation of the design systems, which may subsequently fail to work in some other environments. Therefore, a large range of channel models and channel emulators have been proposed over literature, each specific for a particular scenario [2].

While communicating over an acoustic channel, small differences in velocities of the sound waves lead to preferred propagation directions for the transmitted signal owing to the fluctuations in water temperature, salinity, and density. A collection of transmission rays travels along each of the preferred directions and are referred to as path bundles. Such assortments of communication paths are reflected and delayed both at the base and the surface of the waterbody [3].

The acoustic channel is sparse in nature and exhibits multipath behavior along with long propagation delay owing to the low traveling speed of sound. Multipath components are also associated with spreading around the specular components and diffused reflections contributing to the fading behavior. The propagation speed in turn is affected by temperature, pressure, salinity, soil composition, etc. All these effects come together to form a two-dimensional waveguide through which acoustic waves can travel for hundreds of kilometers [4].

3.2 Acoustic Propagation Phenomenon

Acoustic waves propagate through a medium by means of adiabatic compression and relaxation. Acoustic waves are characterized by acoustic pressure, particle velocity, particle displacement, and acoustic intensity. Acoustic pressure in one dimension is characterized by acoustic wave equation,

$$\frac{\delta^2 p}{\delta x^2} - \frac{1}{c}\frac{\delta^2 p}{\delta t^2} = 0 \tag{3.1}$$

where p is the acoustic pressure in Pascal, x is the position in the direction of propagation of waves in meters, c is the speed of sound in meter/second, and t is the time in seconds. The solution to (3.1) for any lossless medium is given by D'Alembert's equation [],

$$p = R\cos(\omega t - kx) + (1 - R)\cos(\omega t + kx) \tag{3.2}$$

where ω is the angular frequency in radians/second, t is time in seconds, k is the wavenumber in radians/meter, and R is the unitless traveling wave coefficient. $R = 1$ for waves traveling to right and $R = 0$ for wave traveling to the left. Standing wave is obtained at $R = 0.5$. The propagation speed $\bar{c}$ depends on the medium of propagation and is given by Newton-Laplace equation,

$$\bar{c} = \sqrt{C/\rho} \tag{3.3}$$

where C is the stiffness constant or the bulk modulus and ρ is the density in Kg/m^3 of the medium.

3.2.1 Acoustic Waves for Underground Communications

Acoustic wave communications that have recently gained popularity are different applications enabling communications in underground scenarios like smart seismic exploration, earthquake monitoring, buried pipeline monitoring, and smart drilling for oil and gas reservoirs [5, 6].

Propagation path loss through soil for acoustic waves can be universally described by the equation

$$\mathcal{L}_s = \rho e_s \tau c_m p \tag{3.4}$$

where ρ is the lost due to rain factor, e_s is the loss factor due to erosion, tau is the topographic factor of the soil, c_m is the cover management, and p is the support practices factor.

As acoustic propagation modeling is still very much application-specific let us first consider the oil and gas drilling application. In such an application, a magnetostrictive actuator is used to convert electrical signals into acoustic vibrations. The acoustic signals propagate over the drill string to the bottom and back up to the surface via geophones. A series of alternating short and long resonators is used to model the drill string. Each of the resonators is described by a scattering matrix $\mathbf{S}$. The string is modeled as a two-port device with $\mathbf{S}_{11}$ and $\mathbf{S}_{22}$ representing reflection $\mathbf{S}_{12}$ and $\mathbf{S}_{21}$ representing transmission. Here,

$$\mathbf{S}_{11} = \mathbf{S}_{22} = \hat{r}\left(1 - \frac{(1 - \hat{r}^2)e^{-2j\gamma l}}{1 - \hat{r}^2 e^{-2j\gamma \hat{r}}}\right) \tag{3.5}$$

and

$$\mathbf{S}_{12} = \mathbf{S}_{21} = 1 - \frac{(1 - \hat{r}^2)e^{-2j\gamma l}}{1 - \hat{r}^2 e^{-2j\gamma l}} \tag{3.6}$$

where l is the length of the segment in a string, $\hat{r}$ is the reflection coefficient, γ is a function of the attenuation coefficient α, f is the frequency of operation, and v is the velocity of the acoustic wave, as, $\gamma = \frac{2\pi f}{v} - j\alpha$. Here, a sub-matrix $\mathbf{T}$ is separately defined as

$$\mathbf{T} = \frac{1}{\mathbf{S}_{12}}\begin{bmatrix} \mathbf{S}_{12}\mathbf{S}_{21} & \mathbf{S}_{11} \\ -\mathbf{S}_{22} & 1 \end{bmatrix} \tag{3.7}$$

In this case, the channel frequency response of the whole string $\mathbf{T}_s$ can then be obtained as

$$\mathbf{T}_s = \prod_{i=1}^{N} \mathbf{T}_i \tag{3.8}$$

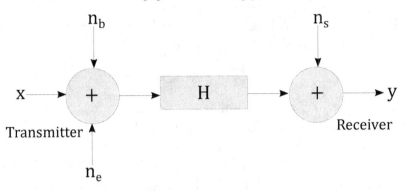

FIGURE 3.1
Equivalent underground acoustic communication model of the drill string.

where $i = 1, 2, \ldots, N$ is the number of segments of the string. The equivalent model of a drill string is presented in Fig. 3.1. Let x be the transmitted signal, y is the received signal, $\mathbf{H}$ is the channel frequency transfer function, n_b is the drill bit noise, n_e is the environmental noise, and n_s is the surface noise. The received signal y in this case can be given by

$$y = \mathbf{H}x + \mathbf{H}(n_b + n_e) + n_s = \mathbf{H}(x + n_d) + n_s \qquad (3.9)$$

where n_d is the total transmitter noise.

3.2.2 Acoustic Waves for Underwater Communications

When underwater objects vibrate, they put the water molecules around them in motion consistently and the later molecules flow in step with the former ones. This is how sound pressure waves are created. The water molecules alternately compress and relax as the acoustic waves travel through the sea. The change in the density of seawater helps acoustic waves travel through sea [7].

Due two different temperature, pressure, and salinity of water, the velocity of acoustic waves can be affected leading to preferred propagation direction after transmission. A simple formula for the three independent factors controlling the velocity of acoustic waves was given by Medwyn in 1975,

$$C = 1449.2 + 4.66T - 0.055T^2 + 0.00029T^3$$
$$+ (1.34 - 0.010T)(S - 35) + 0.016z \qquad (3.10)$$

where T is the temperature in degrees centigrade, S is the salinity in parts per thousand of dissolved weight of salt, and z is the depth in meters.

3.2.2.1 The Doubly Spread Channel

In the signal processing domain, this results in a channel, which exhibits an impulse response consisting of multiple delayed echoes of the transmitted signal. These echoes are associated with major propagation paths. Additionally, we observe that this multipath channel picture is not a static one. That is, the height, phase, and even the position of the different echoes changes with time, and changes quite rapidly with frequencies in the range of 5–10 Hz. This creates the behavior of the so called *doubly spread channel*, which shows variation in the delay domain, as well as in the time domain.

3.2.2.2 Absorption Loss

The primary mechanism of signal loss results from the conversion of the energy in the propagating signal into heat. This mechanism is referred to as absorption loss. In sea water, the absorption loss of acoustic signals is strongly frequency dependent and increases with increasing frequency [8]. Signal energy decay due to absorption loss is proportional to $e^{a(f)l}$ where $a(f)$ is an increasing function of frequency and l is the distance from the source. In this case, $a(f)$ can be calculated using [8] as

$$a(f) = \frac{0.0106 f_1 f^2}{f_1^2 + f^2} e^{\frac{pH-8}{0.56}} + \left(1 + \frac{T}{43}\right)\left(\frac{S}{35}\right)\frac{0.52 f_2 f^2}{f_2^2 + f^2} e^{-\frac{D}{6}}$$
$$+ 0.00049 f^2 e^{-\left(\frac{T}{27} + \frac{D}{17}\right)} \text{ dB/Km} \qquad (3.11)$$

where $f_1 = 0.78\sqrt{S/35}\, e^{\frac{T}{26}}$ KHz is the relaxation frequency contributed by absorption due to boric acid, $f_2 = 42\, e^{\frac{T}{17}}$ KHz is the relaxation frequency contributed by absorption due to magnesium sulfate, T is the temperature in Celsius, D is the depth in km, and S is the salinity in parts per trillion. The pH value has been considered to be fixed at 8. The practical impact of the frequency dependence of the absorption loss is that the communication channel is effectively band-limited and the available bandwidth is a decreasing function of range. This characteristic can significantly impact the choice of modulation and multi-access techniques as well as the problem of optimizing network topology.

3.2.2.3   Spreading Loss

Spreading losses are caused due to the reflection/refraction of the fixed amount of transmitted energy by the sound speed fluctuation as the signal propagates away from its source. Two key factors contributing to spreading losses are multipath propagation and formation of shadow zones.

3.2.2.4   Multipath Propagation

In most environments and at the frequencies of interest for communication signals, the ocean can be modeled as a waveguide with a reflecting surface and ocean bottom and a spatially varying sound speed in the water. The reflections of acoustic signals from the sea surface and bottom and the refraction of signals by the spatially varying sound speed in the water column result in multiple propagation paths from each source to receiver. This multipath results in a delay spread in the often time-varying impulse response of the communications channel leading to intersymbol interference at the receiver. The delay spread of this impulse response can be significant at times ranging up to 100 millisecond, as reported in [9, 10].

The temporal fluctuations in the channel impulse response can be driven by both time variations in the propagation environment and motion of the transmitting or receiving platforms. Environmental variation can give rise to rapid temporal fluctuations (i.e., Doppler spread) in the channel resulting in the challenge of estimating the parameters of a rapidly fluctuating system (i.e., the channel impulse response) with an apparently large number of independent parameters (i.e., samples of the impulse response). Therefore, the multipath phenomenon results in too many challenges, each of which needs close examination depending on different underwater environments. Therefore we will investigate that in detail in the next part, where we will classify the underwater acoustic channel.

With symbol rates of up to 5000 symbols/second common in modern phase coherent systems, delay spreads result in intersymbol interference that can extend for 100s of symbols. For high rate phase coherent systems, the receiver must either explicitly or implicitly estimate this impulse response in order to successfully estimate the data sequence that has been transmitted through the channel. The ability of the receiver to do this depends upon the delay spread and rate of fluctuation of the channel impulse response and is a primary factor in determining the capability of the channel to support such communications.

3.2.2.5 Shadow Zones

The refraction of signals by the sound speed fluctuation not only gives rise to multipath but can result in the formation of "shadow zones" [11]. These are areas where there is little propagating signal energy. Thus it could be difficult to communicate with a receiver located in a shadow zone. Shadow zones can occur both in deep water (1000s of meters) and longer ranges (10s of kilometers) as well as in shallow water (up to 100 meter) and shorter ranges (upto 3 kms). In these environments, the vertical movement of masses of water results in vertical movement of the sound speed structure of the water column. This phenomenon gives rise to variations in the location of shadow zones, even for the case of a stationary source and receiver.

In order to characterize shadow zones as a function of depth and transmission range, let us consider a source close to the surface. The rays transmitted upwards near the horizontal are rapidly refracted downwards, because of the shallow negative gradient.

The limit ray, tangential to the surface, is emitted upwards at angle β_0, $\cos\beta_0 = c_0/c_S$. The ray bundle transmitted in the angular sector $[0, \beta_0]$ crossed a certain specified depth at angles spanning the angular sector $[\arccos c_A/c_0, \arccos c_A/c_S]$, which is narrower than the transmitted sector. These grouped rays all follow approximately the same refracted geometrical path and therefore they undergo a very small divergence spreading. The same ray bundle can be followed after its refraction by the deep gradient, concentrated in a convergence zone. The empty space left by this ray concentration is called a shadow zone.

Variations in received signal-to-noise ratios (SNRs) by as much as 10 dB on time scales of several hours were observed in shadow zones [12] and shown to dramatically impact communications system performance. The temporal fluctuation in the location of regions of low received signal levels impacts the planning of network topologies and adjustment of message routing as the quality of the channel between source/receiver pairs slowly changes.

3.2.2.6 Scattering Loss

One of the most challenging communication scenarios is encountered in the presence of the scattering of some of the transmitted signal by the moving sea surface. The rough sea surface gives rise to a spreading in delay of each surface bounce path, can reduce the spatial correlation of scattered signals, and can result in very high intensity and rapidly fluctuating arrivals in the channel impulse response. Two major phenomena resulting in scattering losses are sea-surface scattering and formation of bubbles at the sea-surface.

3.2.2.7 Surface Scattering

When the sea surface is calm, each surface scattered path results in an arrival in the impulse response that is both fairly stable and localized in delay. In such cases, the impulse response of the channel is often sparse (i.e., has significant arrivals at only a few locations in delay). As the surface becomes more dynamic and roughens, the arrivals not only begin fluctuating in time but also become spread in delay. This results in the need to track a more rapidly varying and less sparse impulse response.

External scattering events that can cause fairly abrupt failures of communication links are referred to as "surface wave focusing," which result in very high intensity and rapidly fluctuating arrivals. Surface wave focusing results from the fact that waves moving over the sea surface can act as downwardly facing curved mirrors that reflect the sound down into the water column and focus it at predictable locations [13]. The surface scattering loss can be approximately calculated as

$$S_L = 10 \log_{10} \left(1.516 \times 10^{-4} f^{3/2} h^{8/5} \gamma\right) \text{ dB} \qquad (3.12)$$

where f is the frequency of operation in Hz, h is the rms surface roughness height in m, and γ is the direction cosine contributed by the incident wave and is calculated using the following expression: $\gamma = 1 + 125\, e^{-2.64(1-1.75)^2 - 50\cot^2\theta}$, where θ is the grazing angle in degrees.

The role of surface scattering in determining communications link quality can result in the link quality having a periodic characteristic when the surface waves are nearly periodic. Knowledge of or the ability to reasonably predict the periodic nature of the quality of a particular communications link would be instrumental in improving transmit scheduling, selecting error correction coding and interleaving strategies, and improving message routing in underwater acoustic communications networks.

3.2.2.8 Bubbles

Bubbles formed by breaking waves at the sea-surface have a major effect on high frequency acoustic propagation in both the open ocean and regions near the shore. Layers of bubbles near the surface can result in a significant attenuation of surface-scattered signals. For bubble densities, characteristic of wind speeds up to around 6 m/s no bubble-induced losses were reported. Above this level, bubble-induced losses increased as a function of wind speed with almost total signal loss (approximately 20 dB loss per surface bounce) at wind speeds of approximately 10 m/s.

The surface scattering loss encountered due to the formation of air bubbles can be calculated approximately using the following empirical expression

$$S_S = 10 \log_{10} \left(10^{-5.05} (1+v)^2 (f+0.1)^{v/150} \tan^\beta \theta \right) \text{ dB/m}^2 \quad (3.13)$$

with $\beta = 4\left(\frac{v+2}{v+1}\right) + (2.5(f+0.1)^{-1/3} - 4) \cos^{1/8} \theta$, where f is the frequency of operation in KHz, v is the wind speed in knots, and θ is the grazing angle in degrees.

Bubble clouds injected down into the water column also significantly attenuate propagating signals with rates as high at 26 dB/m being reported [14]. The injection of bubbles by a breaking wave in shallow water can result in a sudden channel outage.

3.2.2.9   Impact on Signal Power

The overall signal power loss over a distance l for a signal of frequency f in the underwater acoustic channel is given by

$$A(l, f) = A_0 l^k a(f)^l \quad (3.14)$$

where A_0 is a unit normalizing constant, k is the spreading factor (it is typically around 1.5) accounting for spreading losses, and $a(f)$ is the frequency-dependent absorption coefficient given by (3.11). The normalized approximation of (3.14) is given by

$$A(l, f) = l^k 10^{\frac{l}{10^4} \left(0.11 \frac{f^2}{1+f^2} + 44 \frac{f^2}{4100+f^2} + 2.75*10^{-4} f^2 + 0.003 \right)} \quad (3.15)$$

where the normalization is done such that $A(l = 1\text{m}, f) = 1$, which we will need below.

3.2.2.10   Ambient Noise

In order to obtain values for the signal-to-noise ratio (SNR), which is the ultimate determinant of the channel capacity, we need to factor in the different sources of underwater ambient noise, primarily turbulence, shipping, surface gravitational wave, and thermal receiver noise. The power spectral density of each noise source has been studied empirically in [15], and the respective equations are given in (3.16) with units of μPa per Hz as a function of f in kHz [15]. The total contribution of the noise is the sum of the different sources, given in (3.16). Each noise source dominates different portions of the spectrum. At frequencies below 10 Hz turbulence dominates; for frequencies between 10 and 100 Hz shipping activity becomes dominant and is modeled by the shipping parameter s (where $0 < s < 1$). For frequencies ranging from 100 Hz to 100 kHz surface noise due to gravitational waves and determined by the wind speed, w_s is the dominant noise source, and at frequencies above 100 kHz, thermal noise becomes the most important factor. The power spectral density of the different noise sources is given as[1]

$$10 \log_{10} N_t(f) = 107 - 30 \log f$$
$$10 \log_{10} N_s(f) = 40 + 20(s - 0.5) - 26 \log(f) - 60 \log(1/f + 0.03)$$
$$10 \log_{10} N_w(f) = 50 + 7.5\sqrt{w_s} - 20 \log(f) - 40 \log(100/f + 0.4) + 40$$
$$10 \log_{10} N_{th}(f) = -15 + 20 \log f - 60.$$

$$(3.16)$$

where N_t is the noise contributed by turbulence, N_s is the noise contributed by shipping activity, N_w is the noise contributed by the surface gravitational waves, w_s is the wind speed, and N_{th} is the noise contributed by thermal noise at the receiver. The total noise power spectral density is then simply

$$N(f) = N_t(f) + N_s(f) + N_w(f) + N_{th}(f). \qquad (3.17)$$

3.3 Modeling Underwater Acoustic Links

As mentioned before there are different ways of modeling the underwater acoustic propagation link depending on the scenario region and range of application and type of water body dealt with.

---

[1]Note these equations are copied erroneously from [15] in many publications we have consulted.

3.3.1 Sea Surface

Acoustic waves are reflected and phase shifted by 180° off the sea surface resulting in destructive multipath interference. The power spectral density (PSD) of the acoustic signal experiencing attenuation due to rough sea surface and fully developed sea winds can be computed as

$$S_{PM}(k) = \frac{\alpha}{2k^3} e^{-\beta\left(\frac{g}{k}\right)^2 \frac{1}{U^4}} \qquad (3.18)$$

where $\alpha = 0.0081, \beta = 0.74$ are empirically derived, $g = 9.82$ m/s^2 is the acceleration due to gravity, U is the wind speed at 19.5 meters above the sea surface, and $k = 2\pi/\lambda$ is the angular spatial frequency in radians/meter.

3.3.2 Bathymetry

Bathymetry is a technique used to model the multipath scattering pattern caused by bottom of the sea. This technique is used to model the interaction between the acoustic waves and the ocean bed sediments. The waves get attenuated, phase-shifted, absorbed, and reflected by the sediment layer and the measure of attenuation, absorption, and reflection experienced depends on the grain quality, density, porosity, and composition of the ocean bottom sediments. Bathymetry data is available through global databases like British Oceanographic Data Center [16]. However the data granularity is not consistent everywhere. In many areas of the sea, detailed bathymetric data is not yet available and the variation in ocean floor characteristics is not well captured.

A generic sinusoidal sea-floor topology can be generated using the function,

$$z(x) = R(x) \times \frac{z_{\max}}{2} \left(\sin\left(-\frac{\pi}{2} + \frac{2\pi x}{L_{\text{hill}}} \right) + 1 \right) \qquad (3.19)$$

where $z(x)$ is the elevation at position x, $z_{\max}$ is the maximum hill elevation, L_{hill} is the distance between two adjacent peaks, and $R(x) \in (0, 1]$ is the scaling function that generates uniform random number in different ranges but remains constant between two adjacent peaks.

3.3.3 Beam Modeling

Typically underwater acoustic multipath propagation can be modeled using two types of beam.

- Geometric—Linear boundaries are used to separate the neighboring beams at the point of departure. The rays, that are recorded are taken into consideration, are only the ones that have hat-shaped boundaries and are traveling very close to the receiver. BELLHOP is the most common software used for ray-tracing based on geometric beams. It models rays that arrive and travel very close to the receiver.

- Gaussian—Gaussian intensity profile in a direction normal to the ray is used to model the energy spread over every beam. Superposition of multiple Gaussian beams close to the receiver is used to estimate the total acoustic intensity arriving at the receiver.

3.4 Channel Emulators

The propagation of acoustic waves in any medium is given by (3.1). Fourier transform can be used to move (3.1) from time domain to frequency domain and back using

$$\hat{p}(\omega) = \int_{-\infty}^{\infty} p(t)e^{i\omega t}\mathrm{d}t \qquad (3.20)$$

and

$$p(t) = \frac{1}{2\pi} \int_{-\infty}^{\infty} \hat{p}(\omega)e^{-i\omega t}\mathrm{d}t \qquad (3.21)$$

where $\omega = 2\pi f$ with f as the frequency of operation. Here $\hat{p}(\omega)$ is the complex acoustic pressure that can be used to solve the three dimensional Helmholtz equation

$$\frac{\delta^2 p}{\delta x^2} + \frac{\omega^2}{c^2}\hat{p} = 0 \qquad (3.22)$$

where $\omega/c = k$ is the wave number. There are different ways of solving (3.1). However the procedure followed to arrive at the solution depends on the kind of water-body/medium depth and salinity of the medium and the range of transmission or the distance between the transmitter and the receiver. The most generalized way of finding the solution is "ray tracing." Ray tracing, though location-specific, provides a complete physical understanding of the propagation phenomenon at high frequency.

3.4.1 Ray tracing for underwater channels

Rays can be used to represent sound emitted from a source while propagating in different directions. Sum of the contributions from these rays is used to calculate the sound pressure field at the receiver. The sound pressure field existing between the source and the receiver are referred to as eigenrays. The eigenrays follow straight lines if the medium is homogeneous. The rays are refracted on the way if the medium is heterogeneous and the velocity of the acoustic waves varies with distance.

3.4.1.1 Ray Trajectories

Theoretically, the ray solution is formed by seeking an asymptotic series of the form

$$\hat{\wp}(r, z) \sim e^{i\omega\tau(r,z)} \sum_{p=0}^{\infty} \frac{h_p(r, j)}{(i\omega)^p} \tag{3.23}$$

where $\hat{\wp}(r, z)$ is the solution to the 2-D Helmholtz equation $\nabla^2\hat{\wp} + \frac{\omega^2}{c^2(\mathbf{x})}\hat{\wp} = -\partial(\mathbf{x} - \mathbf{x}_0)$, where in Cartesian coordinates $\mathbf{x} = (r, z)$, $c(\mathbf{x})$ is the sound speed and ω is the angular frequency of the source located at $\mathbf{x}_0$. To obtain the ray equations, we seek a solution to the 2-D Helmholtz equation, in the form of (3.23), where (3.23) is called the *ray series*. The ray series is generally divergent, but in some cases it can be shown to be an asymptotic approximation to the exact solution. Taking derivatives of the ray series in (3.23) only with respect to the x-coordinate, we obtain $\hat{\wp}_r = e^{i\omega\tau}\left[i\omega\tau_r \sum_{p=0}^{\infty} \frac{h_p}{(i\omega)^p} + \sum_{p=0}^{\infty} \frac{h_{p,r}}{(i\omega)^p}\right]$ and, $\hat{\wp}_{rr} = e^{i\omega\tau}\left\{\left[-\omega^2\tau_r^2 + i\omega\tau_{rr}\right] \sum_{p=0}^{\infty} \frac{h_p}{(i\omega)^p} + 2i\omega\tau_r \sum_{p=0}^{\infty} \frac{h_{p,r}}{(i\omega)^p} + \sum_{p=0}^{\infty} \frac{h_{p,rr}}{(i\omega)^p}\right\}$.

Thus we can write, $\nabla^2 \hat{\wp} = e^{i\omega\tau}\{[-\omega^2|\nabla\tau|^2 + i\omega\nabla^2\tau]\sum_{p=0}^{\infty}\frac{h_p}{(i\omega)^p} + 2i\omega\nabla\tau\cdot\sum_{p=0}^{\infty}\frac{\nabla h_p}{(i\omega)^p} + \sum_{p=0}^{\infty}\frac{\nabla^2 h_p}{(i\omega)^p}\}$. Substituting this result into the Helmholtz equation and equating terms of like order in ω, we obtain the following infinite sequence of equations for the functions $\tau(\mathbf{x})$ and $h_p(\mathbf{x})$,

$$\mathcal{O}(\omega^2) : |\nabla\tau|^2 = c^{-2}(\mathbf{x})$$
$$\mathcal{O}(\omega) : 2\nabla\tau\cdot\nabla h_0 + (\nabla^2\tau)h_0 = 0$$
$$\mathcal{O}(\omega^{1-p}) : 2\nabla\tau\cdot\nabla h_p + (\nabla^2\tau)h_p = -\nabla^2 h_{p-1}, \qquad p = 1, 2, \ldots$$

$$(3.24)$$

The $\mathcal{O}(\omega^2)$ equation for $\tau(\mathbf{x})$ is known as the *eikonal equation*. The remaining equations for $h_p(\boldsymbol{x})$ are known as the *transport equations*.

3.4.1.2 Solving the Eikonal and the Transport Equations

The eikonal equation

$$|\nabla\tau|^2 = \frac{1}{c^2(\boldsymbol{x})} \qquad (3.25)$$

is a first-order non-linear partial differential equation (PDE) that can be solved by a variety of equations. The most convenient of these is to use the method of characteristics [17]. In this case, a family of curves (rays) is introduced, which are perpendicular to the level curves (wavefronts) of $\tau(\boldsymbol{x})$. This family of rays defines a new coordinate system, and it turns out that in ray coordinates the eikonal equation reduces to a far simpler, linear ordinary differential equation. Since $\nabla\tau$ is a vector perpendicular to the wavefronts, we can define the ray trajectory $\boldsymbol{x}(s)$ by the following differential equation $\frac{d\boldsymbol{x}}{ds} = c\nabla\tau$. The factor of c is introduced so that the tangent vector $\frac{d\boldsymbol{x}}{ds}$ has unit length. Now from the eikonal equation (3.25), the term on the right is found to be unity. Since $\left|\frac{d\boldsymbol{x}}{ds}\right| = 1$, the parameter s is simply the arclength along the ray.

The rays can also be conveniently parameterized with respect to travel time or any other quantity, which increases monotonically along the ray. Our definition for the rays is based on their perpendicular to the level curves of $\tau(\mathbf{x})$, a function which for the moment is still unknown. We consider first just the x-component of the differential equation. Differentiating with respect to s and using the eikonal equation (3.25), we obtain, $\frac{d}{ds}\left(\frac{1}{c}\frac{dr}{ds}\right) = -\frac{1}{c^2}\frac{\partial c}{\partial r}$. By applying this process to each of the coordinates, we obtain the following vector equation for the ray trajectories, $\frac{d}{ds}\left(\frac{1}{c}\frac{d\boldsymbol{x}}{ds}\right) = -\frac{1}{c^2}\nabla c$.

Therefore, in cylindrical coordinates, these ray equations may be written in the first-order form

$$\frac{dr}{ds} = c\xi(s), \qquad \frac{d\xi}{ds} = -\frac{1}{c^2}\frac{\partial c}{\partial r}, \qquad (3.26)$$

$$\frac{dz}{ds} = c\zeta(s), \qquad \frac{d\zeta}{ds} = -\frac{1}{c^2}\frac{\partial c}{\partial z} \qquad (3.27)$$

where $[r(s), z(s)]$ is the trajectory of the ray in the range-depth plane. The auxiliary variables $\xi(s)$ and $\zeta(s)$ are introduced in order to write the equations in the first-order form. The tangent vector to a curve $[r(s), z(s)]$ parameterized by arclength is given by $\left[\frac{dr}{ds}, \frac{dz}{ds}\right]$. Thus from the above equations, the tangent vector to the ray is, $\boldsymbol{t}_{\text{ray}} = c\big[\xi(s), \zeta(s)\big]$ and the normal vector to the ray is, $\boldsymbol{n}_{\text{ray}} = c\big[-\zeta(s), \xi(s)\big]$, where $\boldsymbol{t}_{\text{ray}} \cdot \boldsymbol{n}_{\text{ray}} = 0$. To complete the specification of the rays, the initial conditions are also needed. The initial conditions are that the ray starts at the source position (r_0, z_0) with a specified take-off angle θ_0. Thus we have, $r = r_0, \xi = \frac{\cos\theta_0}{c(0)}, z = z_0, \zeta = \frac{\sin\theta_0}{c(0)}$. If the index of refraction is independent of frequency, then the ray paths are also independent of frequency.

The travel time is obtained by solving the eikonal equation in the coordinate system of the rays such that $\nabla\tau \cdot \nabla\tau = \frac{1}{c^2}$. Therefore, $\frac{d\tau}{ds} = \frac{1}{c}$, which can be solved to obtain

$$\tau(s) = \tau(0) + \int_0^s \frac{1}{c(s)}ds \qquad (3.28)$$

and can be calculated along the curve $[r(s), z(s)]$. Using $\tau(s)$, the path delays over each eigenray arriving at the receiver can be calculated relative to the direct path.

The final step in computing the pressure is to associate an amplitude with each ray and then to solve the transport equation,

$$2\nabla\tau \cdot \nabla h_0 + (\nabla^2\tau)h_0 = 0. \tag{3.29}$$

Consider an arbitrary field F and arbitrary volume V. Using Gauss's theorem, we can state that the integral of the divergence of F throughout the volume is equal to its flux through the boundary of that same volume, $\int_V \nabla \cdot F dV = \int_{\partial V} F \cdot n dS$, where n is an upward pointing normal. From this case, we can conclude that $\int_{\partial V} h_0^2 \nabla\tau \cdot n dS = 0$. Let us define a ray tube as the volume enclosed by a family of rays. Since the rays are normal to the phase fronts, $\nabla\tau \cdot n$ vanishes on the sides of the ray tube. Hence, from the energy conservation law, $\int_{\partial V_0} \frac{h_0^2}{c} dS = \int_{\partial V_1} \frac{h_0^2}{c} dS = \text{const}$, where ∂V_0 and ∂V_1 denote the end caps of the ray tube. Intuitively, this implies a rising or falling amplitudes as the ray tube shrinks or expands. In particular, if we let the ray tube become infinitesimally small and use values at an arbitrary $s = 0$ as a reference, we conclude that,

$$h_0(s) = h_0(0)\left|\frac{c(s)J(0)}{c(0)J(s)}\right|^{1/2} \tag{3.30}$$

where $J(s)$ is an quantity proportional to the cross-sectional area of the ray tube. Next, let us consider a ray tube formed by the boundary of two rays launched with adjacent take-off angles, separated by $d\theta_0$, as shown in Fig. 3.2. The cross-sectional area is just the hypotenuse $Jd\theta$ with, $J = r\left[\left(\frac{\partial z}{\partial\theta_0}\right)^2 + \left(\frac{\partial r}{\partial\theta_0}\right)^2\right]$.

3.4.1.3 Transmission Loss

Each eigenray makes a contribution to the complex pressure field based on its intensity and travel time at that point. The pressure field at the receiver can be calculated by simply summing up the contributions of each of the eigenrays leading to

$$\hat{\wp}(r, z) = \sum_{p=1}^{P(r,z)} \hat{\wp}_p(r, z), \tag{3.31}$$

where P denotes the number of eigenrays contributing to the field at a particular receiver position (r, z) and $\hat{\wp}_p(r, z)$ is the pressure

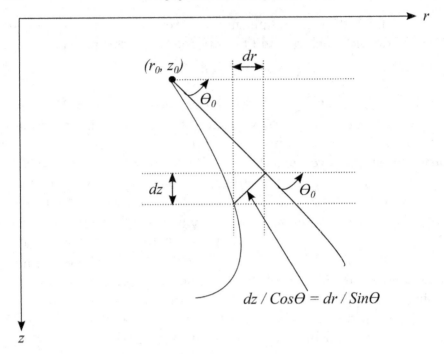

FIGURE 3.2
The ray-tube cross-section.

due to the pth eigenray. The number of contributing eigenrays can vary considerably. In the near field, there may be just three important eigenrays : a direct ray, a bottom-bounce ray, and a surface-bounce ray. The remaining eigenrays strike the bottom with steeper angles than the critical angle and therefore strongly attenuated. At longer ranges, there often will be contributions from paths which strike the surface and bottom several times or eigenrays following different refracted paths. Once we have associated a travel time and intensity with the ray paths, we can complete the calculation of the transmission loss at the receiver as

$$\mathrm{TL}(r, z) = -20 \log \left| \frac{\hat{\wp}(r, z)}{\hat{\wp}^0(s = 1)} \right| \tag{3.32}$$

where $\hat{\wp}^0(s)$ is the pressure for a point source in free space to be evaluated at a distance of 1 m from the source. Thus, $\hat{\wp}^0(s = 1) = \frac{1}{4\pi}$.

3.4.2 Gaussian Beam Modeling

Gaussian beam spreading can be used to model variationsin the propagation environment due to processes such as tides, currents, moving transmitter/receiver/surface, and internal waves. The first step is to model a linear time-varying (LTV) system. Using LTV systems, the multipath channel over an UWAC link can be characterized in time domain using

$$h(\tau, t, f) = \sum_{p=1}^{P} h_p \gamma_p(f, t) \delta(\tau - \tau_p(t)) \qquad (3.33)$$

where P is the total number of distinct echoes or path bundles encountered, with amplitudes h_p and delays τ_p. The fading parameter $\gamma_p(f, t)$ models the small-scale time variations of each path bundle. If we assume that $\gamma_p(f, t)$ is a stationary random process in t, we can derive its statistical description by the time-correlation function. In that case, the channel Doppler spectrum can be modeled using the Fourier transform of the auto-correlation function of the fading parameter, $\gamma_p(f, t)$

$$R_p(f, \Delta t) = \mathrm{E}\left[\gamma_p(f, t)\gamma_p^*(f, t + \Delta t)\right] \qquad (3.34)$$

where $*$ denotes the complex conjugate. The channel Doppler spectrum is approximated by a stretched exponential distribution

$$S_p(f, \nu) = \mathcal{F}\left(R_p(f, \Delta t)\right) \propto e^{-\left(\frac{|\nu|}{\alpha}\right)^{\beta}} \qquad (3.35)$$

where $\mathcal{F}(\cdot)$ denotes Discrete Fourier Transform (DFT), ν are the Doppler frequencies, α is the parameter that matches the Doppler power spectrum to the exponential function, and β is the stretching exponent. There is no closed-form expression for $R_p(f, \Delta t)$ and $\mathcal{F}(R_p(f, \Delta t))$ is not strictly exponential. Therefore, we cannot exactly generate the stretched exponential spectrum with an auto-regressive memory process in the time domain, as done in [18]. Hence, in Subsection 3.4.2.3, we resort to the SOS model for generating random processes directly from the Doppler spectrum.

3.4.2.1 WSSUS Model

Assuming stationarity and uncorrelated echoes with infinite bandwidth, the delay power spectrum of the channel can be given by

$R_h(\tau) = \mathrm{E}[h(t,\tau)h^*(t,\tau')] = \sum_{p=1}^{P} \mathrm{E}\left[|h_p(t)|^2\right] \delta(\tau - \tau_p)$. On the other hand, the frequency correlation function can be computed as $\mathrm{E}\left[\mathcal{F}\{h(t,\tau')\}\mathcal{F}\{h^*(t,\tau)\}\right] = \frac{1}{2\pi}\sum_{p=1}^{P} \mathrm{E}\left[|h_p(t)|^2\right]e^{j2\pi(f-f')\tau_p}$. We note that the frequency correlation function depends only on the difference $f - f'$. This means that the uncorrelated scattering assumption induces wide-sense stationarity in the frequency domain. The time variation captured in (7.8) in the WSSUS case can be computed separately for each delay τ since delays on different path bundles are uncorrelated. In this case, the scattering function in the delay domain can be calculated as

$$S(\tau, \nu) = \mathcal{F}\left\{ \sum_{p=1}^{P} \mathrm{E}\left[h_p(t)h_p^*(t + \Delta t)\right]\delta(\tau - \tau_p) \right\} \qquad (3.36)$$

where the Fourier transform $\mathcal{F}\{\cdot\}$ is computed with respect to Δt. In this case, $S(\tau_p, \nu)$ will describe the time variations in the p-th path around its Doppler spectrum.

3.4.2.2 Discrete-Time Channel Model

A plethora of UWAC channel measurement campaigns like KAM, MACE, and SPACE [18] conducted in different locations revealed that Rician distribution offers precise characterization of the channel stochastic behavior. A variety of Rician K_R factors are encountered for different sets of reflected path bundles due to surface, bottom, surface-bottom, surface-bottom-surface combination, etc. Therefore, in this section, we will be using the SOS model to generate channel samples whose amplitudes follow Rician fading distribution.

3.4.2.3 The SOS Channel Model

A WSSUS channel with a given scattering function can be conveniently generated using the SOS model, where the channel samples $h_p[i] = h_p(iT_s)$ are generated according to

$$h[i] = \frac{1}{\sqrt{M(1+K_R)}}\left[\sum_{m=1}^{M} e^{j(\phi_m + 2\pi f_m iT_s)} + \sqrt{K_R}\, e^{j(\phi_0 + 2\pi f_0 iT_s)} \right]$$

$$(3.37)$$

where T_s is the sample time interval used to represent the continuous time variable t, M is the number of multipaths to model each channel tap, ϕ_m are random phases uniformly distributed in the interval $[0, 2\pi]$ with initial phase ϕ_0 and f_m are the M Doppler frequencies contributing to each channel tap. In (3.37), K_R is the Rician factor given by the ratio of the power of the specular component to that of the diffused components in each channel tap and f_0 is the Doppler frequency of the specular component in each channel tap[1].

In order to generate (3.37), we need to choose M Doppler frequencies f_m according to the Jakes spectrum, which is actually the Fourier transform of (7.9) and can be given by

$$S(\nu) = \frac{1}{2\alpha} e^{-(|\nu|/\alpha)^\beta} \tag{3.38}$$

where (3.38) serves as the probability density function (PDF) from which the values of f_m can be drawn. We can do this using the *inverse transform sampling lemma*, which states that the output *cumulative distribution function (CDF)* $F(X)$ is distributed uniformly in $[0, 1]$, given that the input X is chosen according to $f(x)$. Therefore we can generate samples following $f(x)$ by choosing y uniform in $[0, 1]$, and applying $x = F^{-1}(y)$. Integrating (3.38), we find the CDF of the Doppler frequencies as

$$F(f_m) = \int_{-\infty}^{f_m} \frac{1}{2\alpha} e^{-(|\nu|/\alpha)^\beta} d\nu = \frac{1}{2\beta} \Gamma\left[\frac{1}{\beta}, -\left(-\frac{f_m}{\alpha}\right)^\beta\right] \tag{3.39}$$

where $\Gamma(\cdot, \cdot)$ is the upper incomplete Gamma function.

Inverse transform sampling allows us to generate an arbitrary probability distribution from a uniform distribution. In this case,

---

[1] If $M \to \infty$, it can be shown that the statistics of (3.37) are exactly identical to those of (3.36), and that the amplitude is Rician distributed. If $K_R = 0$, the amplitude will be Rayleigh distributed. In practice, of course, for $M < \infty$, the model becomes, strictly speaking, non-stationary. However, even for a moderate value of M, a good approximation can be achieved as long as $M \geq 100$.

letting $F(f_m) = u$ to be a uniform distribution in $[0, 1]$, the transformation

$$f_m = -\alpha \left[\sum_{k=0}^{1/\beta-1} (k+1)! \, W \left\{ \frac{1}{k} \left(\frac{(\mathrm{mod}(2u, 1))\beta^2}{\beta - 1} \right)^{1/k} \right\} \right]^{1/\beta}$$

(3.40)

generates Doppler frequencies distributed according to a stretched exponential distribution, "mod" refers to the modulo operation, which returns the remainder after division of $2u$ by 1, W denotes the Product-Log or Lambert-W function.

3.4.2.4 The Tapped Delay Line Representation

In order to build a discrete model for frequency-selective channels, the channel impulse response is modeled as

$$h(t; \tau) = \sum_{p=1}^{P} c_p(t)\delta(\tau - \tau_p)$$

(3.41)

where $c_p(t)$ are the fading coefficients. However, the model in (3.41) is incomplete, as it has infinite bandwidth (revealed by use of delta function). So to use it for system implementation, transmit and receive filtering are to be considered.

Let us consider a communication system where x_i complex transmit samples enter a pulse shaping transmit filter with impulse response $g(t)$, then the channel, and are finally received by a matched receive filter. The output signal $z(t)$ is sampled at multiples of the symbol time T, or multiples of some convenient sample time T_s. The overall structure is a discrete linear time-varying filter and can be implemented using a tapped delay line model, where $l = 1, 2, \ldots L$ are the indices of the filter taps considered for the tapped delay line model. In order to accomplish that, we need to generate the values of the filter taps, $f_l(kT_s)$. In conjunction with the channel model in (3.41), this requires a digital filtering operation in addition to a sampling rate adequate to model the useful bandwidth of the channel (3.41).

The generation of the filter taps is a natural extension of the time-varying path weights (3.37) and can be accomplished by

$$
f_l[k] = \sum_{p=1}^{P} \sqrt{\frac{P_p}{M(1+K_R)}} \left[\sum_{m=1}^{M} e^{j(\phi_{m,p}+2\pi f_{m,p}(kT_s-\tau_p))} \right.
$$

$$
\times\, g(lT_s - \tau_p) + \sqrt{K_R}\, e^{j(\phi_{0,p}+2\pi f_{0,p}(kT_s-\tau_0))} g(lT_s - \tau_0) \Bigg] \quad (3.42)
$$

where τ_0 is the initial delay and $\{P_p\}$ are the average power levels in the P path bundles.

Moreover, if the terminals move at a velocity of v away from each other, a time period t is expanded into a period of $t + vt/c = (1+a)t$, where a is the Mach number and $c \sim 1500$ m/s is the speed of sound in water. In that case, (3.42) can be modified to accommodate for the Doppler shift,

$$
f_l[k] = \sum_{p=1}^{P} \sqrt{\frac{P_p}{M(1+K_R)}} \left[\sum_{m=1}^{M} e^{j(\phi_{m,p}+2\pi f_{m,p}(akT_s-\tau_p))} g((k-l+ak) \right.
$$

$$
\times\, T_s - \tau_p) \Bigg] + e^{j(\phi_{0,p}+2\pi f_{0,p}(akT_s-\tau_0))} \sqrt{K_R}\, g((k-l+ak)T_s - \tau_0) \Bigg].
$$

$$
(3.43)
$$

It is to be noted here that while the generation model of (3.43) is stationary in the limit $M \to \infty$, it is not strictly so for finite M. Keeping M small, however, is paramount in building an emulator with reasonable complexity.

Note on Stationarity

Analysis of UWAC channel impulse responses gathered during sea trials conducted in numerous measurement campaigns has shown that underwater communication channels can hardly ever be modeled as a WSSUS stochastic process. However, even in case of a non-WSSUS channel, it is possible to determine the period of time (*Stationary Time*) and frequency (*Stationary Bandwidth*) range, in which the WSSUS assumption is locally satisfied [19]. If we define $R_h(t, \Delta\tau)$ as the autocorrelation function of $h(t, \tau)$ in time domain,

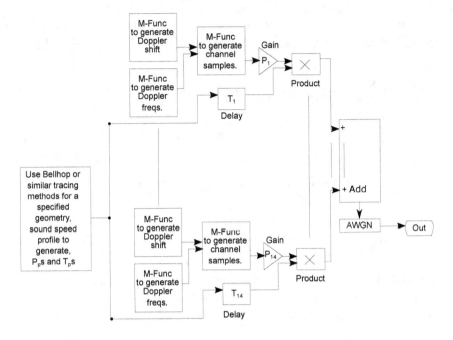

FIGURE 3.3

Simulink-based SOS channel simulation model

then *Stationary Time* can be defined as the maximum observation time interval T_D, for which $R_h(t, \Delta\tau)$ is stationary and the mean value of $h(t, \tau)$ is constant. Consequently, if we define $R_p(f, \Delta t)$ as the autocorrelation function of $\gamma_p(f, \Delta t)$, then *Stationary Bandwidth* can be defined as the maximum frequency range B_D, in which $R_p(f, \Delta t)$ is stationary and mean value of $\gamma_p(f, \Delta t)$ is constant.

3.4.2.5   Simulink Model

The SOS channel representation in (3.43) can be modeled using Simulink. The Simulink model is shown in Fig. 3.3 for a Rician (or Rayleigh) distributed channel, which can represent the impulse response of up to a maximum of 15 channel taps at any discrete time delay (the taps are assumed to be non-overlapping). Separate Matlab functions have been used to generate Doppler shifts and Doppler frequencies. Matlab functions are also included in the

Simulink model in Fig. 3.3 to characterize the time-varying gain $c_p(t)$ in (3.41) with a fixed programmable τ_p.

Fig. 3.4(a) shows sample trajectories generated over a 3-second interval with the Simulink model and $M = 1000$, together with the sample amplitude and Doppler frequency distributions using parameters tabulated in Table 3.1. These parameters are adopted from the ones used for distribution fitting to the measurement campaigns conducted in [18, 20].

The proposed channel emulator is developed using the Simulink platform and is created using the Simulink block-set elements. This emulator can be used to run simulations in Simulink, which can also be converted to RTL code, IP catalog items, or even bit-streams for JTAG co-simulations. This provides an added advantage over prevalent UWAC channel simulators [18, 19, 20] as this Simulink-based emulator can be directly ported to the FPGA platform for hardware implementation. Our emulator can take in a message of symbols and prepare it via upsampling, filtering, and upconversion and then finally downconvert, filter, and downsample the signal after being received over an emulated UWAC link. An example plot for channel impulse response generated using the proposed channel emulator is presented in Fig. 3.4(b).

TABLE 3.1

Parameter Set for Emulation

K_R	α (Hz)	β
3	0.024, 0.44	0.39, 0.63
5	1.27, 3.08	0.50, 1.16

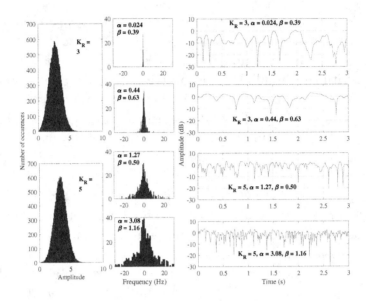

(a) Sample trajectory of the channel emulator with the right-hand side showing a 3-*s* synthesized path gain, the left-hand side showing the Doppler frequency and amplitude distributions.

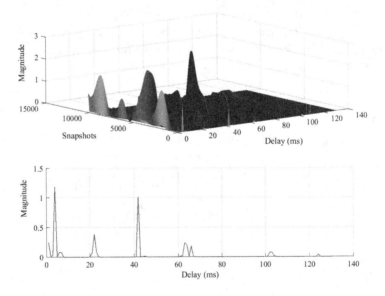

(b) Generated CIR with the top one showing response for 10 channel taps and the bottom one presenting the evolution of a single channel tap.

FIGURE 3.4

Simulink-based channel emulator output.

Bibliography

[1] A. Alimohammad, S. Fouladi Fard, B. F. Cockburn, and C. Schlegel, "A compact and accurate Gaussian variate generator," *IEEE Trans. Very Large Scale Integration Sys.*, vol. 16, no. 5, 2008.

[2] S. F. Cotter and B. D. Rao, "Sparse channel estimation via matching pursuit with application to equalization," *IEEE Trans. on Communications*, vol. 50, no. 3, pp. 374-377, 2002.

[3] P. Hoeher, "A statistical discrete-time model for the WSSUS multipath channel," *IEEE T. Vehicular Tech.*, vol. 41, no. 4, pp. 461–468, 1992.

[4] J. G. Proakis, *Digital communications*, McGraw-Hill, Boston, 2007, ID: 43526842.

[5] M. Stojanovic, "On the Relationship Between Capacity and Distance in and Underwater Acoustic Communication Channel", *First ACM International Workshop on Underwater Networks*, 2006.

[6] R. A. van Walree, T. Jenserud, and R. Otnes, "Stretched-exponential Doppler spectra in underwater acoustic communication channels," *J. Acoust. Soc. Am.*, vol. 128, no. 5, November 2010.

[7] F. Jensen, W. Kuperman, M. Porter, and H. Schmidt, "Computational Ocean Acoustics," (Springer, New York, NY, 2000), pp. 11–12 and 52-54.

[8] C. S. Clay, and H. Medwin, *Acoustical Oceanography: Principles and Applications* (John Wiley & Sons, New York, NY, 1977), pp. 88 and 98–99.

[9] D. Kilfoyle, and A. B. Baggeroer, "The State of the Art in Underwater Acoustic Telemetry," *IEEE J. Oceanic Engineering*, 25(1) pp. 4–27, (2000).

[10] M. Porter, et al., "The Kauai Experiment", in *High Frequency Ocean Acoustics, Eds.*, American Institute of Physics, 2004, pp. 307–321.

[11] M. Siderius, M. Porter, and the KauaiEx Group, "Impact of Thermocline Variability on Underwater Acoustic Communications: Results from KauaiEx", in *High Frequency Ocean Acoustics*, American Institute of Physics, 2004, pp. 358–365.

[12] J. Preisig, and G. Deane, "Surface wave focusing and acoustic communications in the surf zone," *J. Acoust. Soc. Am.*, vol. 116(4), pp. 2067–2080, (2004).

[13] D. M. Farmer, G. B. Deane, and S. Vagle, "The influence of bubble clouds on acoustic propagation in the surf zone," *IEEE J. Oceanic Engineering*, 26(1) pp. 113–124, (2001).

[14] M. A. Ainslie, and J. G. McColm, "A simplified formula for viscous and chemical absorption in sea water", *Journal of the Acoustical Society of America*, 103(3), pp. 1671–1672, 1998.

[15] R. Coates, *Underwater Acoustic Systems*, New York: Springer, 1982.

[16] Url: https://www.bodc.ac.uk/

[17] N. Bleistein, *Mathematical Methods for Wave Phenomena* (Academic, Orlando, FL, 1984).

[18] P. Qarabaqi, and M. Stojanovic, "Statistical Characterization and Computationally Efficient Modeling of a Class of Underwater Acoustic Communication Channels," *IEEE J. of Oceanic Engg.*, vol. 38, no. 4, pp. 701-717, Oct. 2013.

[19] Y. Zhu, S. Le, L. Pu, X. Lu, Z. Peng, J.-H. Cui, and M. Zuba, "AquaNet Mate: A real-time virtual channel/modem simulator for Aqua-Net," in *Proc. MTS/IEEE Ocean*, Bergen, Norway, Jun. 2013.

[20] M. Chitre, "A high-frequency warm shallow water acoustic communications channel model and measurements," *J. Acoust. Soc. Amer.*, vol. 122, no. 5, pp. 2580-2586, Nov. 2007.

4

Magneto-Inductive Propagation

CONTENTS

This chapter details out a few theoretical models that are used to characterize propagation of magneto-inductive (MI) waves in different environments. MI waves are preferred modes of communication for underground scenarios. MI-wave based communication system uses wired coils to exchange information guided by Faraday's electromagnetic law of induction. Faraday's law estimates how much electromotive force (EMF) is generated due to interaction between a magnetic field and an electric circuit. The EMF generated is directly proportional to the induced voltage, the change in magnetic flux and the change in time. A modulated current flowing through the transmitter induces a time-varying magnetic field in space between the transmitter and the receiver, which in turn, induces sinusoidal current in the receiver. The induced current in receiver is demodulated to recover the transmitted information.

DOI: 10.1201/9781003213017-4

4.1   Introduction

Magneto-inductive Communication (MIC) uses wired coils to exchange information guided by the Faraday's electromagnetic law of induction [1]. Faraday's law estimates how much Electromotive Force (EMF) is generated due to interaction between a magnetic field and then electrical circuit. The EMF is given by, $\epsilon = -N\frac{\Delta\phi}{\Delta t}$, where ϵ is the induced voltage, N is the number of loops in the wired coils, $\Delta\phi$ is the change in magnetic flux, and Δt is the change in time. For an MIC system, a modulated current flowing through the transmit coil (TC) induces a time-varying magnetic field in space between the transmitter and the receiver, which in turn, induces sinusoidal current in the receiver coil (RC). The induced current in RC is demodulated to recover the transmitted information.

Though MIC suffer from high propagation path loss over long-distance links and cannot be used for transferring high data rates, MIC offers several advantages over traditional EM and acoustic wave-based communications especially in unconventional environments like underwater (sea-water), underground (soil, rock), and the interface between different media (soil and air or water and air) [2, 3, 4, 5, 6, 7, 8, 9, 10]. The benefits worth mentioning are:

- Stable Channel Response—A very small portion of the radiated energy escapes to the far field as the resistance of the coil is smaller than the electric dipole. Doppler shifts are mitigated by the high velocity of the propagating field. Moreover magnetic permeability of the medium (water, earth) stays the same even with time, location, and dimensions. Therefore the resultant channel responses are stable and predictable.

- Negligible Propagation Delay—MI wave can penetrate water and solid mediums with the speed of light way much faster than EM waves or acoustic waves. This enables near-zero latency in delivery of information in underwater or underground scenarios.

- Low-cost, Low-power—Use of small coil and a very little

amount of energy required to drive those coils makes MIC a low-cost low-power option. It is an energy efficient form of communication especially for short distance links in underwater and underground scenarios.

Based on the above-mentioned advantages, MIC emerges as a promising candidate for some specific application scenarios like

- Leakage detection in oil and gas pipelines in both underwater and underground situations

- Timely and robust detection of pre-disaster scenarios in underwater and underground like tsunami typhoons undersea earthquakes

- Military applications that require in audibility and invisibility of the transmitted signal underwater and underground

- Monitoring physical chemical and biological states of the sea through an underwater wireless sensor network

- Monitoring of underground structures like mines buried sites pipelines through wireless underground sensor networks

All these applications are possible due to the fact that media like solid soil and water have almost similar magnetic permeabilities and the variation in their properties result in little variation in the attenuation rate of the magnetic fields.

4.2   The Propagation Phenomenon

Fig. 4.1 depicts the block diagram of a point-to-point MIC system. The induced magnetic field between the TC and the RC is modulated by the waveform carrying information. RC is never equipped with its own power source and the magnetic field coupling is the strongest along the coil axis. The channel or the space over which the magnetic field is induced between the TC and the RC can be modeled in two different ways; the *equivalent circuit analysis* and the *magnetic field analysis*.

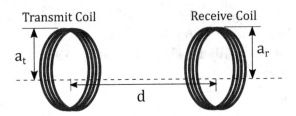

FIGURE 4.1

A point-to-point MIC System.

4.2.1   Magnetic Field Analysis

The transmitter is modeled as a magnetic dipole loop antenna driven by alternating current. Maxwell's equation is used to describe the generated magnetic dipole field and the field analysis is used to derive the coupling between the transceiver coils. The functions of the involved parameters are used to derive the MI path loss [11, 12, 13, 14].

4.2.1.1   MI Path loss Model in Lossless Medium

The strength of the induced magnetic field decreases with the increase in distance between the transmitter and the receiver. The mutual inductance between a pair of coupled coils is given by

$$\mathcal{M} = \frac{\mu \pi N_t N_r a_t^2 a_r^2}{2\sqrt{(a_t^2 + d^2)^3}} \quad (4.1)$$

where μ is the magnetic permeability of sea water, N_t is the number of turns in the transmit coil, N_r is the number of turns in the receive coil, a_t is the radius of the transmit coil, a_r is the radius of the receiver coil, and d is the distance between the transmitter and the receiver. From Fig. 4.1, it is evident that in order to maximize the received power, the load impedance Z_L is chosen to be equal to the complex conjugate of the output impedance of the secondary loop and can be expressed as

$$Z_L = R_r + \frac{\omega^2 M^2 R_t}{R_t^2 + \omega^2 L_t^2} + j\left(\frac{\omega^2 M^2 R_t}{R_t^2 + \omega^2 L_t^2} + \omega^2 L_t^2 - \omega L_r \right) \quad (4.2)$$

where R_r is the resistance of the receiver coil, ω is the angular frequency of the signal, R_t is the resistance of TC, L_t is the self-inductance of the TC, and $\mathcal{M}$ is the mutual inductance developed. Using the above formulation, the path loss experienced over a lossless medium when an MIC system is used is given by

$$PL_{MI} = -10\log \frac{R_1^2 \omega^2 \mathcal{M}^2}{R_t(R_L + R_r)^2 + R_t(X_L + \omega L_r)^2} \tag{4.3}$$

where R_L and X_L are the real and the imaginary parts of the load impedance Z_L, respectively, and L_r is the self-inductance of the RC.

4.2.1.2 Path loss in Underwater Scenarios

The exponential decrease in signal strength or transmit field strength caused by eddy currents induced by the time-varying magnetic field is given by

$$PL_{UW} = 20\log(e^{d\sqrt{\pi f \mu \sigma}}) \tag{4.4}$$

where $\alpha = \sqrt{\pi f \mu \sigma}$ is the attenuation coefficient of the eddy currents equal to the in, d is the distance between TC and RC, σ is the medium conductivity, μ is the permeability of water and f is the frequency of operation.

Path loss in underwater environment is quantified as the sum of the path losses experienced in lossless medium (PL_{MI}) and the path loss experienced due to eddy currents resulting from that time-varying magnetic field (PL_{UW}), i.e., $\rho_{UW-MI} = PL_{MI} + PL_{UW}$. The skin effect is a phenomenon in which electromagnetic field or electrical current tends to accumulate near the surface of the conductor instead of propagating through the core of the conductor. From (4.4), it is evident that more the water is conductive, higher is the eddy current loss experienced. The only way to alleviate the situation and improve communication in deep underwater scenario is to select low operating frequency for MIC.

4.2.1.3 Path loss in Underground Scenarios

The path loss experienced in underground environment can be calculated as

$$\rho_{UG-MI} = 6.4 + 20\log r + 20\log \beta + 8.69\alpha d \qquad (4.5)$$

where d is the distance between the transmitter and the receiver, α is the attenuation constant, β is the phase shift constant, and their values depend on the dielectric properties of the soil. Values of α and β are calculated using the Peplinski principle, which gives the relative dielectric constant of the soil water mixture as

$$\epsilon_m = \epsilon'_m - j\epsilon''_m \qquad (4.6)$$

where ϵ'_m and ϵ''_m are dielectric related constants and depend on the volume of water fraction, bulk density, and specific density of solid soil particles and soil-type dependent constants based on the fraction of sand and clay present in the soil.

4.2.1.4 MI Noise Model

Underwater and underground MI systems do not experience considerable noise or interference owing to the low number of natural magnetic sources in underground or underwater environment. The few number of sources that are present do not contribute much interference as the propagation field strength decreases considerably with distance. The only source of noise that plagued underground or underwater scenario is the thermal noise or Johnson noise yielded by the random motion of the electrons in the medium given by $N_t \approx BkT$, where B is the communication bandwidth, T is the temperature in Kelvin, and k is the Boltzmann constant given by $k \approx 1.38 \times 10^{-23} J/K$. Therefore lower the water temperature, higher is the noise power.

4.2.1.5 Path loss in Long-range Underwater MIC

Extending MIC over a long range can be realized through multihop transmission networks. The multihop transmission involves deploying multiple relays between the transmitter and the receiver if their relays do not need additional power sources and processing units. The link between the transmitter and the receiver can be viewed as a waveguide (refer to Fig. 4.2) assuming that the transmitter, receiver, and the relay coils all have resistance R, the overall path loss experienced can be given by

$$\rho_{MI-WG} \approx 10\log\frac{4(Z_{n-1,n}+R)}{R} + 20\log\xi\left(\frac{R}{\omega M}, n-1\right) \quad (4.7)$$

where $Z_{n-1,n}$ is the reflected impedance of the nth coil on the $(n-1)$th coil and $\xi\left(\frac{R}{\omega M}, n-1\right)$ is the recursive function composed of $(n-1)$ order polynomials of $\frac{R}{\omega M}$ given by

$$\xi\left(\frac{R}{\omega M}, n-1\right) = b_{n-1}\left(\frac{R}{\omega M}\right)^{n-1} + b_{n-2}\left(\frac{R}{\omega M}\right)^{n-2}$$
$$+ \ldots + b_2\left(\frac{R}{\omega M}\right)^2 + b_1\left(\frac{R}{\omega M}\right) + b_0 \quad (4.8)$$

where $b_i, i = 0, 1, 2, \ldots, n-1$ is the set of polynomial coefficients and n are the total number of nodes presenting the network including the transmitter and the receiver. For the underwater scenario, the overall path loss for transmission over a MI relay-based waveguide is given by

$$\rho_{UW-MI-WG} = \rho_{MI-WG} + \rho_{UW-WG} \quad (4.9)$$

where

$$\rho_{UW-WG} = 20(n-1)\log e^{r\sqrt{\pi f\mu\sigma}} \quad (4.10)$$

where r is the distance between two adjacent communicating nodes. For the underground scenario $R/\omega M$ can be further reduced to

$$\frac{R}{\omega M} = \frac{4R_0}{fN\mu\pi}\left(\frac{r}{a}\right)^3 \quad (4.11)$$

where R_0 is the wire resistance, N is the number of turns in the coil, f is the operating frequency, and a is the radius of the inducting coil.

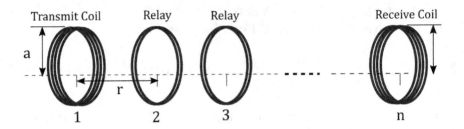

FIGURE 4.2
Multi-hop relay-based Underwater MIC System.

4.2.2   Equivalent Circuit Analysis

In this case the system is modeled either as a transformer or a two-port network. The TC consumes the transmit power and the power consumed across the load impedance is the received power. The channel path loss is calculated by formulating the relation between the transmit and the received power.

4.2.2.1   System Modeling

An MI transmitter-receiver pair can be modeled as the primary and the secondary coils of a transformer. Using equivalent analysis for the transformer we can write

$$Z_t = R_t + j\omega L_t$$

$$Z_{rt} = \frac{\omega^2 M^2}{R_r + j\omega L_r Z_L}$$

$$Z_r = R_r + j\omega L_r$$

$$Z_{tr} = \frac{f^2 M^2}{R_t + j\omega L_t}$$

$$U_M = -j\omega M \frac{U_s}{R_t + jfL_t} \tag{4.12}$$

where U_M is the induced voltage on the receiver coil, Z_{tr} and Z_{rt} are the influences of the receiver on the transmitter and vice versa, respectively, Z_t and Z_r are the self-impedances of the transmitter and the receiver, respectively, with self-inductances L_t and L_r and coil resistances R_t and R_r; $\mathcal{M}$ is the mutual inductance between them, U_S is the voltage of the transmitter's battery and $\omega = 2\pi f$ where f is the operating frequency. The received (P_r) and transmit powers (P_t), in turn, are functions of the transmission range d,

$$P_r(d) = \text{Re}\left\{ \frac{Z_L \cdot U_M^2}{(Z_r' + Z_r + Z_L)^2} \right\} \tag{4.13}$$

and

$$P_t(d) = \text{Re}\left\{ \frac{U_S^2}{Z_t' + Z_t +} \right\} \tag{4.14}$$

where Z_t' and Z_r' are the complex conjugate of the self-impedances Z_t and Z_r of the transmit and the receive coils, respectively. The coil resistances R_t and R_r are given by

$$R_t = N_t 2\pi a_t R_0$$
$$R_r = N_r 2\pi a_r R_0 \tag{4.15}$$

where N_t and N_r are the number of turns of the transmitter and the receiver coils, respectively, a_t and a_r are the radii of the transmit and receive coils, respectively, and R_0 is the resistance of a unit length of the loop. The self-inductances can be derived as

$$L_t \approx \frac{1}{2}\mu\pi N_t^2 a_t$$
$$L_r \approx \frac{1}{2}\mu\pi N_r^2 a_r \tag{4.16}$$

Using Stokes theorem the mutual inductance can be given by, $\mathcal{M}\mu\pi N_t N_r \frac{a_t^2 a_r^2}{2d^3}$, where μ is the permeability of the medium. The system model presented in Fig. 4.2 can be extended to a waveguide model for multihop transmission [15, 16]. Applying equivalent circuit analysis, we can derive

$$Z = R + j\omega L + 1/j\omega C$$

$$Z_{i(i-1)} = \frac{\omega^2 \mathcal{M}^2}{Z + Z_{(i+1)i}} \quad i = 2, 3, \ldots, n-1$$

$$Z_{i(i+1)} = \frac{\omega^2 \mathcal{M}^2}{Z + Z_{(i-2)(i-1)}} \quad i = 3, 4, \ldots, n$$

$$U_{\mathcal{M}_i} = -j\omega \mathcal{M} \frac{U_{\mathcal{M}_{i-1}}}{Z + Z_{(i-2)(i-1)}} \quad i = 2, 3, \ldots, n \text{ and } U_{\mathcal{M}_1} = U_S$$

$$(4.17)$$

where $Z_{i(i-1)}$ is the influence of the ith coil on the $(i-1)$th coil and vice versa; $U_{\mathcal{M}_i}$ is the induced voltage on the earth coil. In this case, the received power can be calculated as

$$P_r = \text{Re}\left\{ \frac{Z_L U_{\mathcal{M}_n}^2}{(Z_{n(n-1)} + Z + Z_L)^2} \right\} \quad (4.18)$$

where Z_L is the load impedance associated with the receiver coil, $\text{Re}\{\cdot\}$ represents the real part of the expression, and $U_{\mathcal{M}_n}$ is the induced voltage on the nth coil.

4.3   Channel Emulator Example

If the transmitter and the receiver are in two different media, for example, one in air and the other in soil or water, the system models need to be developed and in a bit different way. In this case the magnetic intensity generated by the coil is captured through the full-wave model. Fig. 4.3 represents the inter-media MIC system in terms of cylindrical coordinate system where z_1 and z_2 are the distances between the transmitter and the inter-media boundary and the receiver and the inter-media boundary, respectively.

Owing to the symmetric structure of magnetic coils, cylindrical coordinate system can be used to model the time-varying current i_t generated in the coil. The corresponding magnetic field is given

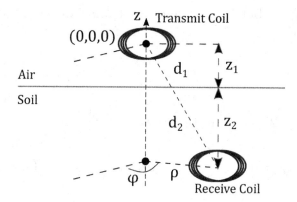

FIGURE 4.3
Inter-media MIC system in terms of cylindrical coordinate system.

by

$$\mathbf{h}_c^x = \left[h_\rho^x, h_\phi^x, h_z^x\right]^T \tag{4.19}$$

where x is the coil radius ($x = a$ for coil above the boundary and $x = u$ for the coil located under the boundary), ρ, ϕ, z are the cylindrical coordinates. Let us also denote n and i as the number of turns and current flowing through the coil, respectively. Also $\mu, epsilon, \sigma, k = \omega\sqrt{\mu\epsilon}$ and η represent the permeability, real permittivity, conductivity, propagation constant, and wave impedance of the medium, respectively, ϵ is the complex permittivity of the medium, $\epsilon = \hat{\epsilon} + j\sigma/\omega$. For the transmitter above the boundary in medium 1,

$$h_z^a = -\frac{ji_t a_t^2 n_t}{8} \int_{-\infty}^{\infty} dk_\rho \frac{k_\rho^3}{k_{1z}} H_0^{(1)}(k_\rho\rho)\gamma(k_{1z}, z) \tag{4.20}$$

where $\gamma(k_{1z}, z) = (\alpha e^{jk_{1z}|z|} + R_{12}e^{jk_{1z}z + 2jk_{1z}z_1})$, $\alpha = 1$, $k_{1z} = \sqrt{k_1^2 - k_\rho^2}$, ρ is the horizontal distance, $H_0^*(x)$ is the Hankel function of the first kind and the zeroth order, and R_{12} is the reflection coefficient at the boundary given by

$$R_{12} = \frac{\mu_2 k_{1z} - \mu_1 k_{2z}}{\mu_2 k_{1z} + \mu_1 k_{2z}} = \frac{k_{1z} - k_{2z}}{k_{1z} + k_{2z}}. \tag{4.21}$$

The relation between h_z and the transverse components (h_ρ^a, h_ϕ^a) is given by

$$[h_\rho, h_\phi] = \frac{1}{k_\rho^2}\left[\frac{\delta}{\delta\rho}\frac{\delta h_z}{\delta z}, \frac{1}{\rho}\frac{\delta}{\delta\phi}\frac{\delta h_z}{\delta z}\right]. \tag{4.22}$$

Putting (4.22) back in (4.19) we can obtain

$$[h_\rho^a, h_\phi^a] = \left[-\frac{i_t a_t^2 n_t}{8}\int_{-\infty}^{\infty} \mathrm{d}k_\rho[k_\rho^2 H_1^{(1)}(k_\rho\rho)]\gamma(k_{1z}, z)\right]. \tag{4.23}$$

Similarly the field below the boundary can be solved. The cylindrical coordinates can be transformed back to the Cartesian coordinate using $\mathbf{h}_r^x = C_1 \mathbf{h}_c^x$, where $C_1 = [\cos\phi - \sin\phi 0; \sin\phi \cos\phi 0; 001]$ is the transformation coefficient matrix. The mutual inductance in this case can be calculated as $\mathcal{M} = \phi/i_t$, where ϕ is the magnetic flux.

Bibliography

[1] C. A. Balanis, *Antenna Theory: Analysis and Design.* Hoboken, NJ, USA: Wiley, 2016..

[2] Y. Morag, N. Tal, Y. Leviatan, and Y. Levron, "Channel capacity of magnetic communication in a general medium incorporating full-wave analysis and high-frequency effects," *IEEE Trans. Antennas Propag.*, vol. 67, no. 6, pp. 4104–4118, Jun. 2019.

[3] H. Guo, Z. Sun, and P. Wang, "Multiple frequency band channel modeling and analysis for magnetic induction communication in practical underwater environments," *IEEE Trans. Veh. Technol.*, vol. 66, no. 8, pp. 6619–6632, Aug. 2017.

[4] I. F. Akyildiz, P. Wang, and Z. Sun, "Realizing underwater communication through magnetic induction," *IEEE Commun. Mag.*, vol. 53, no. 11, pp. 42–48, Nov. 2015.

[5] B. Gulbahar and O. B. Akan, "A communication theoretical modeling and analysis of underwater magneto-inductive wireless channels," *IEEE Trans. Wireless Commun.*, vol. 11, no. 9, pp. 3326–3334, Sep. 2012.

[6] A. K. Sharma et al., "Magnetic induction-based non-conventional media communications: A review," *IEEE Sensors J.*, vol. 17, no. 4, pp. 926–940, Feb. 2017.

[7] Y. Li, S. Wang, C. Jin, Y. Zhang, and T. Jiang, "A survey of underwater magnetic induction communications: Fundamental issues, recent advances, and challenges", *IEEE Commun. Surveys Tuts.*, vol. 21, no. 3, pp. 2466–2487, 3rd Quart., 2019.

[8] R. K. Gulati, A. Pal, and K. Kant, "Experimental evaluation of a near-field magnetic induction based communication system," in *Proc. IEEE Wireless Commun. Netw. Conf. (WCNC)*, 2019, pp. 1–6.

[9] H. Ma et al., "Antenna optimization for decode-and-forward relay in magnetic induction communications," *IEEE Trans. Veh. Technol.*, vol. 69, no. 3, pp. 3449–3453, Mar. 2020.

[10] S.-C. Lin, A. A. Alshehri, P. Wang, and I. F. Akyildiz, "Magnetic induction-based localization in randomly deployed wireless underground sensor networks," *IEEE Internet Things J.*, vol. 4, no. 5, pp. 1454–1465, Oct. 2017.

[11] S. Kisseleff, W. Gerstacker, R. Schober, Z. Sun, and I. F. Akyildiz, "Channel capacity of magnetic induction based wireless underground sensor networks under practical constraints," in *Proc. IEEE WCNC*, Shanghai, China, Apr. 2013, pp. 2603–2608.

[12] B. Gulbahar and O. B. Akan, "A communication theoretical modeling and analysis of underwater magneto-inductive wireless channels," *IEEE Trans. Wireless Commun.*, vol. 11, no. 9, pp. 3326–3334, Sep. 2012.

[13] H. Guo and Z. Sun, "M^2I communication: From theoretical modeling to practical design," in *Proc. IEEE ICC*, Kuala Lumpur, Malaysia, May 2016, pp. 1–6.

[14] Z. Sun and B. Zhu, "Channel and energy analysis on magnetic induction-based wireless sensor networks in oil reservoirs," in *Proc. IEEE ICC*, Budapest, Hungary, Jun. 2013, pp. 1748–1752.

[15] E. Shamonina, V. Kalinin, K. H. Ringhofer, and L. Solymar, "Magneto-inductive waveguide," *Electron. Lett.*, vol. 38, no. 8, pp. 371–373, Apr. 2002.

[16] R. R. A. Syms, I. R. Young, and L. Solymar, "Low-loss magneto-inductive waveguides," *J. Phys. D Appl. Phys.*, vol. 39, no. 18, pp. 3945–3951, Sep. 2006.

5

Optical Wave Propagation

CONTENTS

Light waves or optical waves can also be used to transmit information wirelessly. The propagation of an optical wave is governed by Maxwell's equations. The propagation characteristics depend on the optical property and the physical structure of the medium.

DOI: 10.1201/9781003213017-5

They also depend on the makeup of the optical wave, such as its frequency content and its temporal characteristics. This chapter discusses the basic propagation characteristics of optical waves through a wireless medium. Basic impact of the environment on the propagation of optical waves and different modeling techniques proposed over literature are detailed in consecutive sections of this chapter.

5.1 Introduction

Optical wave communication (OWC) actually precedes radio frequency-based wireless technology. The earliest format of OWC is the use of fire or light to send warnings, announce news, and direct ships or armies. They have evolved since then into different forms depending on different regions of the EM spectrum that are used for communication; infrared (IR), 10–380 nano-meter(nm), visible light (VL), 380–780 nm and ultraviolet (UV), 780–10^6 nm. For example, for the IR case, IR laser diodes or light-emitting diodes (LEDs) are used as transmitters. In VL communications (VLC), phosphor-based white LEDS or green, red, blue LEDs are used as transmitters while positive-intrinsic-negative (PIN) diode, avalanche photodiode, or camera are used as receivers. Wireless UV communications (WUVC) techniques are mainly used for long-range communications with transmitters like carbon arcs, low pressure mercury arc lamps, gallium lamps, and nitrogen-filled tubes and receivers like photo-multiplier tubes (PMT) with different operating frequencies, and the propagation channel between the optical transmitter and the receiver behaves differently [1, 2, 3, 4].

Optical frequencies (light waves) constitute the highest frequency range in the electromagnetic spectrum ranging between 0.3 and 30,000 Terahertz (THz). This means that the wavelength range is shorter than radio waves varying between 10 and 10^6 nm. Owing to shorter wavelengths, optical waves suffer high propagation path loss ($\sim 1/\lambda^2$) and scattering ($\sim 1/\lambda^4$). Therefore in a wireless environment, the range of successful communication using optical wave depends on the wavelength of operation. The longest wavelength falls in the IR range, and therefore they are used for long-distance communications in indoor, outdoor, and indoor-to-outdoor environments; coined as wireless infrared communications (WIRCs). VLC works with short-range wavelength and therefore used for short-range communications in indoor, outdoor, underwater, and underground scenarios; coined as VLCs. UV spectrum covers the lowest possible wavelength and therefore can be used for very small distance communication; coined as WUVCs [5, 6, 7, 8, 9, 10].

Other than wavelengths of operation, it is possible to classify OWC based on the presence or absence of line-of-sight (LoS) path between the transmitter and the receiver, degree of directionality, divergence angle, and field-of-view between them. The four most common scenarios encountered through this classification are: a) Direct LoS, b) Non-direct LoS, c) Non-direct non-LoS (NLOS) and d) user mobility tracking. The first one is used for point-to-point and peer-to-peer networking with the transmitter being oriented toward the receiver. The second one is suitable for point-to-multipoint indoor scenarios with and without wide LoS beam between the transmitter and receiver. The third one is good for diffuse link configuration for indoor WIRCs and outdoor WUCs scenarios with no LoS between the transmitter and the receiver. The fourth one is capable of tracking user mobility within an indoor scenario for observing the receiver within a specific area and aim the optical beam close or parallel to the required path. An overall schematic diagram of an OWC system is provided in Fig. 5.1.

5.2 Propagation Phenomenon

Whatever be the communication scenario, the channel characteristics of OWC depend on two factors; a) type of communication environment and b) the positions and the mutual orientation of

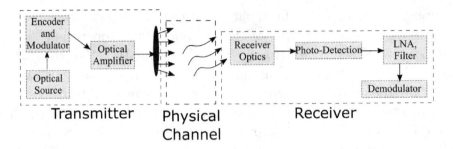

FIGURE 5.1

Overall schematic of an end-to-end OWC system.

the transmitter and the receiver and their orientation toward reflecting surfaces. The major advantages of OWC channel is that it does not experience multipath fading, frequency offset and Doppler frequency shift as opposed to traditional radio frequency-based communication channel. However the major drawback in OWC channels is the short distance of reliable communication and multipath dispersion resulting in excruciating orders of intersymbol interference (ISI).

5.2.1 General Characteristics

The fundamental parameters used to model propagation over an OWC channel are summarized below [11, 12, 13, 14],

5.2.1.1 Channel Impulse Response

It is the time evolution of the signal received by the receiver in response to the infinitely short optical pulse transmitted from the optical transmitter. It is denoted by $h(t)$. The overall system can be modeled as a linear time-invariant (LTI) system with input signal $x(t)$ and output $y(t)$, like

$$y(t) = R_\lambda x(t) \otimes h(t) + n(t) \qquad (5.1)$$

where R_λ is the transmit beam radiation pattern at λ as the wavelength of operation, $\otimes$ denotes convolution, and $n(t)$ is the background light noise modeled as additive white Gaussian noise (AWGN).

5.2.1.2   Zero-frequency channel gain

It is the fraction of the transmit power detected at the receiver. The zero frequency response can be calculated as $H(0) = \int_{-\infty}^{\infty} h(t)dt$ where $H(0)$ is also sometimes referred to ask channel DC gain, as $P_r = P_t H(0)$ where P_t and P_r are the average transmit and received powers, respectively. Once P_t and P_r are known, it is possible to calculate the channel signal to noise ratio (SNR) as,

$$SNR = \frac{R^2 H^2(0) P_r^2}{R_b N_0} \tag{5.2}$$

where R is the responsivity of the receiver or detector, N_0 is the noise spectral density, and R_b is the achievable bitrate.

5.2.1.3   Root mean-squared (RMS) delay spread

It describes the temporal and spatial dispersion suffered by the transmitted signal on its way to the receiver. The RMS delay spread is calculated using

$$D_{rms} = \sqrt{\frac{\int_{-\infty}^{\infty}(t-\mu_\tau)^2 h^2(t)dt}{\int_{-\infty}^{\infty} h^2(t)dt}} \tag{5.3}$$

where t is the propagation time and μ_τ is the mean excess delay given by

$$\mu_\tau = \frac{\int_{-\infty}^{\infty} th^2(t)dt}{\int_{-\infty}^{\infty} h^2(t)dt} \tag{5.4}$$

from which the channel coherence bandwidth can be calculated as $B_c = 1/5D_{rms}$. The RMS delay spread value can be used to calculate the maximum possible bitrate that can be transmitted as $R_b \leq 1/10D_{rms}$. If the data is transmitted at rates higher than R_b, the transmit signal will be crippled with ISI.

5.2.1.4   Frequency Response

It is the effect of the environment on the transmitted signal in frequency domain and is calculated using Fourier transform of the channel impulse response (CIR) $H(f) = \mathcal{FT}\{h(t)\}$. The 3-dB frequency of the channel is calculated using $|H(f_{3dB})|^2 = 0.5|H(0)|^2$.

It is worth mentioning here that lower frequencies are impacted more by higher order reflection than higher frequencies. Low path loss is experienced over highly reflective surrounding resulting in additive higher received power but with higher dispersion.

5.2.1.5 Optical Path loss

It is the overall loss experienced by large-scale fading and shadowing and is very complex in nature affected by environmental parameters like reflectivity, absorption, scattering, and locations of the transmitter and the receiver. The overall path loss can be calculated as

$$PL(dB) = -10 \log_{10} \left(\int_{-\infty}^{\infty} h(t) \right) \tag{5.5}$$

where again the core necessary parameter is the CIR $h(t)$. Therefore the main focus for any OWC channel model is to characterize the CIR accurately.

5.3 Channel Modeling

In order to model different communication scenarios and OW propagation for different wavelengths, several efforts have been done over the years and several environment-specific models have been proposed using either detailed measurement campaign or statistical empirical studies. Channel models proposed so far over literature can be broadly classified into two groups: deterministic and stochastic models [15, 16, 17, 18, 19, 20].

5.3.1 Deterministic Models

If detailed description of specific environment, transmitter and receiver position, configuration and orientation are included into the channel model, that group of channel models are classified as *deterministic*.

5.3.1.1 Recursive Model

The recursive model starts by characterizing the radiation intensity pattern $R(\phi)$ using a Lambertian function

$$R(\phi) = \frac{m+1}{2\pi} \cos^m(\phi), \quad \phi in[-\pi/2, \pi/2] \qquad (5.6)$$

where ϕ is the angle of irradiance or departure and m is the mode number from the transmitter of the radiation lobe. If the receiver is modeled as an active area A_R collecting radiation at an angle θ of the incident rays, the received signal will be proportional to $A_R \cos(\theta)$. In a recursive model, the CIR $h(t)$ is formulated by superposition of the direct LoS component and the infinite sum of multiple reflected components and can be expressed as

$$h(t; \mathbf{S}, \mathbf{R}) = h^0(t; \mathbf{S}, \mathbf{R}) + \sum_{k=1}^{K} h^k(t; \mathbf{S}, \mathbf{R}) \qquad (5.7)$$

where $\mathbf{S}$ and $\mathbf{R}$ are the optical source and the receiver, respectively present within a room scattered with Lambertian reflectors and $h^k(t)$ is the CIR of the components experiencing k-fold reflections with $h^0(t)$ as the discrete LoS component expressed as

$$h^0(t) \approx \frac{(m+1)}{2\pi D^2} A_R \cos^m(\phi) \cos(\phi)\delta(t - D/c) \qquad (5.8)$$

where c is the speed of light and D is the distance between the transmitter and the receiver. The main idea behind the recursive model is to break the reflecting surfaces into several small Lambertian reflecting elements, ϵ, each with area ΔA. Each cell acts as an elemental transceiver with ϵ^s and ϵ^r representing the transmitted and the reflected part, respectively. In that case, the CIR with k reflections can be expressed as

$$h^k(t; \mathbf{S}, \mathbf{R}) = \frac{(m+1)}{2\pi} \sum_{i=1}^{N} \frac{\cos^m(\phi) \cos(\theta)}{D^2} \text{rect}(2\theta/\pi)h^{k-1}$$
$$\times (t - D/c; S_i, \mathbf{R})\Delta A \qquad (5.9)$$

where N is the number of cells, ρ is the cell reflectivity coefficient, and S_i is the location of the ith cell. The idea is to calculate the

CIR after k reflections and then calculate the CIR for all possible $k - 1$ reflections. In this way, (5.9) can be applied to any transmitter under any communication scenario. The main disadvantage of the recursive model is that the accuracy in modeling increases with the number of reflections; however considering more number of reflections increases the computational time and complexity exponentially. A better option is the iterative model.

5.3.1.2   Iterative Model

The main difference between recursive and iterative model is that instead of transceiver cells, we assume the transceiver boxes. The LoS CIR can still be calculated using (5.9). In that case, the CIR after k bounce reflections can be given by

$$h^k(t; \mathbf{S}, \mathbf{R}) \approx \sum_{i=1}^{N} \rho \epsilon_i^r h^{k-1}(t; \mathbf{S}, \epsilon_i^r) \otimes h^0(t; \epsilon_i^s, \mathbf{R}) \qquad (5.10)$$

In (5.10), we can put $\mathbf{R} = \epsilon_j^r$ to obtain

$$h^k(t; \mathbf{S}, \epsilon_i^r) = \sum_{j=1}^{N} h^{k-1}(t - D_{ij}/c; \mathbf{S}, \epsilon_j^r) \qquad (5.11)$$

where $\alpha_{ij} = \text{rect}(2\theta/\pi)\frac{\rho \epsilon_j^r \cos^m(\phi_{ij})\cos(\theta_{ij})}{\rho^2 D_{ij}^2}$ where ϕ_{ij} is the angle between the transmitter and the receiver, θ_{ij} is the angle between the receiver and the transmitter, D_{ij} is the distance between the transmitter ϵ_i^S and the ϵ_j^r receiver, and ρ is the cell spatial partitioning factor. Iterative algorithm reduces computational complexity to some extent; however the computation time still remains a problem.

5.3.1.3   DUSTIN Algorithm

Instead of slicing into number of reflections suffered, it is possible to slice the received signal into time steps. The computation complexity reduces manifold from N^k to N^2 where N is the number of reflecting transceiver cells or blocks. This is what is done in DUSTIN algorithm, which is a simulation-based algorithm using mathematical formulations propounded for the iterative and recursive models. DUSTIN approach also suffers from the fact that they are site-specific as the CIR are accumulated in a disc before analyzing and therefore gets outdated once the transmitter or the receiver moves from one environment to another.

5.3.1.4 Ceiling Bounce Model (CBM) Algorithm

This model was proposed specifically for indoor scenarios with co-located transmitter and receiver. The diffused components of the channel are characterized by the CIR as

$$h(t, a) = H(0) \frac{6a^6}{(t+a)^7} u(t) \qquad (5.12)$$

where $u(t)$ is the unit step function, $a = 2H_c/c$ with H_c as the height of the ceiling from the transmitter and the receiver and c is the velocity of light in meter per seconds. The corresponding delay spread can be calculated as $D_{rms} = \frac{a}{12}\sqrt{\frac{13}{11}}$ with 3dB cut-off frequency given by, $f_{3dB} = \frac{0.925}{4\pi D_{rms}}$. For the LoS component, the CIR can be written as,

$$h(t) = \frac{A_R H^2}{\pi D^4} \delta(t - D/c) \qquad (5.13)$$

where H is the vertical distance between the transmitter and the receiver, D is the horizontal distance between them, A_R is the effective area of the receiver collecting the transmitted radiation. Owing to its simplicity, CBM is often used to design communication algorithms for implementations in VLCs.

5.3.1.5 Geometry-Based Deterministic Models (GBDMs)

GBDMS-based on ray tracing used in literature for modeling different indoor scenarios are based on commercial software Zemax. Using sequential or non-sequential ray tracing capabilities, Zemax can characterize the interaction between rays emitted from optical source within a predefined indoor environment. Based on the geometry of the environment, LoS component and a group of reflected rays will arrive at the receiver. The characteristics of the reflected rays depend on the properties of the objects. What happens within a cubic room is presented in Fig. 5.2. However, most reflecting objects can be modeled using Lambert or the Phong models. If there are more specular components within the environment, Phong model is a better fit. According to Phong's model, the illumination of each surface point I_p can be given by

$$I_p = k_a i_a + \sum_{m \in \text{lights}} (k_d(\hat{L}_m.\hat{N})_{im,d} + k_s(\hat{R}_m.\hat{V}\alpha_{im,s})) \qquad (5.14)$$

where k_s, k_d, and k_a are specular, diffused, and ambient reflection constants, respectively, α is the brightness constant for the source material $\hat{L}_m$ is the direction vector from surface to each light source, $\hat{N}$ is the normal to the illumination point on the surface, $\hat{R}_m$ is the direction taken by the perfectly reflected light, and $\hat{V}$ is the direction pointing toward the viewer. Importing data from Zemax, it is possible to obtain the CIR as

$$h(t) = \sum_{i=1}^{N_r} P_i \delta(t - \tau_i) \qquad (5.15)$$

where N_r is the total number of rays received, and P_i and τ_i are the power and propagation time of the ith ray and $\delta(t)$ is the Dirac delta function.

5.3.2 Stochastic Models

If the impulse response of the OWC channel are collected and characterized by a predefined probability distribution in a stochastic fashion, that group of channel models are classified *stochastic*. If the geometry of arrangement of transmitter, receiver, and the scatterers are predefined according to a particular probability distribution, the impulse response is characterized according to laws of electromagnetic wave propagation. Such channel models are referred to as stochastic models.

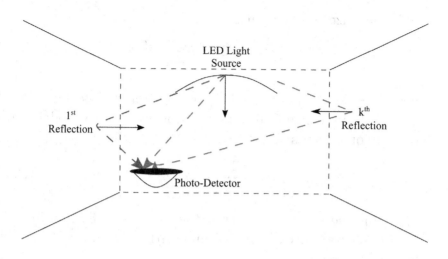

FIGURE 5.2
Cubic room representation of GBDM Modeling scenario.

5.3.2.1 Geometry-based Stochastic Model

Based on integration of sphere photometry, the CIR is obtained by integrating a sphere and then scaled to a specific room as

$$h_{diff}(t) = \frac{H_d}{\tau} e^{-t/\tau} \quad (5.16)$$

representing the diffused component with the decay time τ and H_d is the channel gain of the diffused component given by

$$H_d = \frac{A_R}{A_{\text{room}}} \rho/\langle \rho \rangle \quad (5.17)$$

where A_{room} is the room area, $\langle \rho \rangle = \frac{1}{A_{\text{room}}} \sum_i \Delta A_i \rho_i$ is the average reflectivity of the room and the room is assumed to be divided into several cells each with area ΔA_i and reflectivity ρ_i.

5.3.2.2 Carruther's Model

A combination of geometric model for the environment with iterative technique for calculating multiple reflections form the base of the Carruther's model. Modified Rayleigh distribution is used to characterize the LoS path

$$f(h^0) = \frac{2\alpha\beta^2 h^0 e^{-(\beta h^0)^2}}{1 - \bar{\alpha}e^{-(\beta h^0)^2}} \tag{5.18}$$

where $\bar{\alpha} = 1 - \alpha$ is the shape parameter and β is the scale parameter. Shifted lognormal distribution is used to model diffuse channel components

$$f(h) = \frac{1}{(h - \theta)\sigma\sqrt{2\pi}}e^{-\frac{1}{2\sigma^2}(\ln(h-\theta)-\mu)^2} \tag{5.19}$$

where θ is the shift parameter of the lognormal distribution with mean μ and variance σ^2 and gamma distribution is used to model the delays spread D_{rms}.

5.3.2.3   RS-GBSM Model

If GBSM models are extended to environments where the arrangement of scatterers can be modeled in terms of 2D or 3D shapes, the models are referred to as the RS-GBSMs. In 2D case, the scatterers are assumed to be arranged on a ring or an ellipse. In the 3D case, the schedulers are placed on a sphere or an ellipsoid. RS-GBSMEs are efficient, tractable models that can be easily used for designing different communication techniques for indoor VLC.

5.3.2.4   Monte-Carlo Algorithms (MCA)

MCA is a modification of the DUSTIN algorithm with the assumption of the generalized Lambertian radiation pattern. MCA is implemented through the steps of ray generation, wall processing, and calculation of the photodetector response. The main advantage of MCA is its low computational complexity and simulation time. However, it requires higher number of transmitted rays than the received ones for an accurate mapping on the propagation scenario.

5.3.2.5   Modified Monte-Carlo Algorithms (MMCA)

Modified MCA combines MCA with ray-tracing where Phong's model or Lambertian model is used to model the reflection from the scatterers surrounding the actual path of propagation between the transmitter and the receiver. MMCA is way more accurate and has low computation time; however, it can be location-specific.

5.3.2.6 Hayasaka-Ito Model

The idea is to decompose the CIR into primary $h^1(t)$ and higher order $h^{high}(t)$ reflections where the primary reflection is Gamma distributed with

$$h^1(t) = \frac{\beta^{-\alpha}}{\Gamma(\alpha)} t^{\alpha-1} e^{-t/\beta} \qquad (5.20)$$

where $\Gamma(\alpha)$ is the Gamma function, α and β are the characteristic parameters of the channel. The higher order reflections are modeled using spherical models.

5.4 Additional Factors

There are several factors affecting OWC like absorption, scattering, and scintillation that cause signal degradation and contribute to the atmospheric attenuation. In general, this attenuation is quantified using Beer's law equation

$$\tau = e^{-\beta L} \qquad (5.21)$$

where τ is the atmospheric attenuation, β is the total attenuation coefficient given by $\beta = \beta_{abs}\beta_{scat}$, and L is the distance between the transmitter and the receiver. Scattering and absorption is contributed by the particles suspended in atmosphere (aerosols and molecules). They are of different nature, concentrations, shapes, and sizes. The total absorption due to these particles is given by β_{abs} and the scattering due to them is given by β_{scat}.

5.4.1 Absorption

Collision between the photons within the transmit light beam and the finely suspended particles results in absorption. The amount of absorption can be measured in terms of absorption coefficient given by, $\beta_{\text{abs}} = \alpha_{\text{abs}} N_{\text{abs}}$ where α_{abs} is the effective cross-sectional area of the atmospheric particles and N_{abs} is the concentration of the particles causing absorption within the atmosphere.

5.4.2 Scattering

Physical interaction between the propagating light and the suspended atmospheric particles results in the dispersion of the light beam into a wide range of directions, a phenomenon referred to as scattering. Scattering can be measured by scattering coefficient, $\beta_{\text{scat}} = \alpha_{\text{scat}} N_{\text{scat}}$, where α_{scat} and N_{abs} are the cross-sectional area and concentration of the scattering particles, respectively. Light after scattering is still polarized and of the same wavelength as the transmitted one. Three types of scatterings are encountered: a) Rayleigh, b) Mie and c) Non-selective.

If the particle size is lower than the transmit wavelength, Raleigh scattering is encountered with coefficient, $\beta_m = \alpha_m N_m$ where α_m and N_m are the cross-sectional area and density of the particles with

$$\alpha_m = \frac{8\pi^3 (n^2 - 1)^2}{3N^2 \lambda^4} \tag{5.22}$$

where n is the refractive index, λ is the transmit wavelength, and N is the cubic density of the particles. if the particle diameter is equal to or larger than $1/10$ of the transmit wavelength, Mie scattering is encountered with coefficient $\beta_a = \alpha_a N_a$ where α_a and N_a are cross-sectional area and concentration of the particles.

If the particle dimension is significantly higher than the transmit wavelength, the resultant scattering is a non-selective one. This kind of scattering is often caused by rainfall and the coefficient is given by

$$\beta_{\text{rain}} = \pi a^2 N_a Q_{\text{scat}} \left(\frac{a}{\lambda} \right) \tag{5.23}$$

where a is the radius of the particle, N_a is the cubic concentration of the particles, and Q_{scat} is the scattering efficiency with

$N_a = R/1.33(\pi a^3)V_a$ where R is the rate of rainfall, V_a is the precipitation speed and $V_a = \frac{2a^2 \rho g}{9\eta}$, where ρ is the density of water, g is the gravitational constant (980 cm/sec^2), and η is the viscosity of air (1.8×10^{-4} g/cm.sec) for case when the non-selective scattering is caused by rainfall.

5.4.3 Turbulence

The optical propagation path suffers from random fluctuations in their refractive index due to temperature, pressure, and wind variations both in space and time. The relationship between refractive index (n), pressure (p), and temperature (T) of the medium is given by $n - 1 \approx 79 \times \frac{P}{T}$, where T is in degree Kelvin.

These fluctuations in refractive index result in phase shifts in the propagating optical signals thereby distorting the wave-fronts. These variations in refractive indices are referred to as eddies. Optical waves passing through these eddies are reflected in parts randomly resulting in a distorted wave-front. The light beam bends if the size of the turbulent eddies is larger than the beam diameter. More details on other effects experienced by the optical waves when they traveled a long distance from the satellites or distant ground station, like scintillation, beam spreading, beam wandering, geometric losses will be provided in Chapter 7.

5.5 Underwater Optical Wave Communications (UOWC)

Recent developments in underwater sensor networks and the possibility of communication in the visible light spectrum had sparked research and development of UOWC links. Ocean water is transparent in the visible light spectrum and attenuates light in the blue-green spectral range (450 to 550 nm) within negligible limits. UOWC though works on shorter communication links (100 to 200 nm) offers high throughput, energy efficiency, low latency, and implementation cost.

Propagation of light through water is affected by the physical properties of water, like pressure, temperature, salinity, and the particles suspended in water. Path loss over the LoS path is guided by Beer-Lambert's law

$$PL(\text{optical} - dB) = 10 \log_{10} e^{-K(\lambda)D} \qquad (5.24)$$

where D is the distance between transmitter and the receiver and $K(\lambda)$ is the attenuation coefficient. Light-wave propagation in water is guided by radiative transfer equation (RTE) presented in vector form as

$$\left[\frac{1}{c}\frac{\delta}{\delta t} + \mathbf{n} \cdot \nabla\right] I(t, \mathbf{r}, \mathbf{n}) = \int_{4\pi} \xi(\mathbf{r}, \mathbf{n}, \mathbf{r}, \mathbf{n}') I(\mathbf{r}, \mathbf{n}, \mathbf{r}, \mathbf{n}') \mathrm{d}n' - K(\lambda)$$
$$\times I(t, \mathbf{n}, \mathbf{r}, \mathbf{n}') + E(t, \mathbf{r}, \mathbf{n}') \qquad (5.25)$$

where $\mathbf{n}$ is the direction vector, $\mathbf{r}$ is the position vector, ∇ is the divergence operator with respect to $\mathbf{r}$, I is the irradiance, $\mathbf{E}$ is an internal source radiance and ξ is the volume scattering function (inherent property of water governing light propagation through it varies from water to water).

Channel modeling techniques, detailed so far in this section, like MCA, MMCA, Iterative, Stochastic models and ray-tracing like Zemax can also be applied to model UOWC. Characteristic parameters like CIR ($h(t)$), path loss, delay spreads (D_{rms}), zero-frequency gain ($H(0)$) can be extracted from different modeling techniques for different scenarios like pure sea, clear sea, turbid harbor, coastal ocean, and estuary. An example emulator for modeling movement of light through water has been shared in [21, 22].

Bibliography

[1] C.-X. Wang, F. Haider, X. Gao, X.-H. You, Y. Yang, D. Yuan, H. Aggoune, H. Haas, S. Fletcher, and E. Hepsaydir, "Cellular architecture and key technologies for 5G wireless communication networks," *IEEE Commun. Mag.*, vol. 52, no. 5, pp. 122–130, Feb. 2014.

[2] C.-X. Wang, S. Wu, L. Bai, X. You, J. Wang, and C.-L. I, "Recent advances and future challenges for massive MIMO channel measurements and models," *Sci. China Inf. Sci.*, vol. 59, no. 2, pp. 1–16, Feb. 2016.

[3] A. Gupta and R. K. Jha, "A survey of 5G network: Architecture and emerging technologies," *IEEE Access*, vol. 3, pp. 1206–1232, July 2015.

[4] T. S. Rappaport, S. Sun, R. Mayzus, H. Zhao,Y. Azar, K. Wang, G. N. Wong, J. K. Schulz, M. Samimi, and F. Gutierrez, "Millimeter wave mobile communications for 5G cellular: It will work!," *IEEE Access*, vol. 1, pp. 335–349, May 2013.

[5] M. Tilli, T. Motooka, V-M. Airaksinen, S. Franssila, M. Paulasto-Krockel, and V. Lindroos, 2nd Ed., *Handbook of Silicon Based MEMS Materials and Technologies*, London: William Andrew, 2015.

[6] S. Dimitrov and H. Haas, *Principles of LED light communications towards networked Li-Fi*, London: Cambridge University Press, 2015.

[7] R. J. Drost and B. M. Sadler, "Survey of ultraviolet non-line-of-sight communications," *Journal of Semiconductor Science and Technology*, vol. 29, no. 8, pp. 1–11, June 2014.

[8] Z. Xu and B. M. Sadler, "Ultraviolet communications: Potential and state-of-the-art," *IEEE Commun. Mag.*, vol. 46, no. 5, pp. 67–73, May 2008.

[9] A. R. Young, L. O. Bjorn, J. Moan, and W. Nultsch, 1st Ed., *Environmental UV Photo-biology*, New York: Springer Science, 1993.

[10] M. A. Khalighi and M. Uysal, "Survey on free space optical communication: A communication theory perspective," *IEEE Commun. Surveys and Tuts.*, vol. 16, no. 4, pp. 2231–2258, June 2014.

[11] Z. Ghassemlooy, W. Popoola, and S. Rajbhandari, 1st Ed., *Optical Wireless Communications: System and Channel Modeling with MATLAB*, New York: CRC Press, 2013.

[12] J. M. Kahn, W. J. Krause, and J. B. Carruthers, "Experimental characterization of non-directed indoor infrared channels," *IEEE Trans. Commun.*, vol. 43, no. 2, pp. 1613–1623, Feb. 1995.

[13] K. K. Wong, T. O'Farrell, and M. Kiatweerasakul, "The performance of optical wireless OOK, 2-PPM and spread spectrum under the effects of multipath dispersion and articial light interference," *Int. J. Commun. Syst.*, vol. 13, pp. 551–576, Nov. 2000.

[14] H. Elgala, R. Mesleh, and H. Haas, "Practical considerations for indoor wireless optical system implementation using OFDM," in *Proc. ConTEL'09*, Zagreb, Croatia, June 2009, pp. 25–29.

[15] H. Ding, G. Chen, A. K. Majumdar, B. M. Sadler, and Z. Xu, "Modeling of non-line-of-sight ultraviolet scattering channels for communication," *IEEE J. Sel. Areas Commun.*, vol. 27, no. 9, pp. 1535–1544, Dec. 2009.

[16] R. J. Drost, T. J. Moore, and B. M. Sadler, "UV communications channel modeling incorporating multiple scattering interactions," *JOSA A*, vol. 24, no. 4, pp. 686–695, Apr. 2011.

[17] D. P. Young, J. Brewer, J. Chang, T. Chou, J. Kvam, and M. Pugh, "Diffuse mid-UV communication in the presence of obscurants," in *Proc. ASILOMAR'12*, Pacific Grove, USA, 2012, pp. 1061–1064.

[18] R. Yuan and J. Ma, "Review of ultraviolet non-line-of-sight communication," *China Commun.*, vol. 13, no. 6, pp. 63–75, June 2016.

[19] N. Hayasaka and T. Ito, "Channel modeling of nondirected wireless infrared indoor diffuse link," *Journal of Electronics and Communications in Japan*, vol. 90, no. 6, pp. 10–19, 2007.

[20] H. Hashemi, G. Yun, M. Kavehrad, and F. Behbahani, "Frequency response measurements of the wireless indoor channel at infrared frequencies," in *Proc. IEEE ICCC'94*, New Orleans, USA, 1994, pp. 1511– 1515.

[21] B. M. Cochenour, L. J. Mullen, and A. E. Laux, "Characterization of the beam-spread function for underwater wireless optical communications links," *IEEE J. Ocean. Eng.*, vol. 33, no. 4, pp. 513–521, Oct. 2008.

[22] S. Jaruwatanadilok, "Underwater wireless optical communication channel modeling and performance evaluation using vector radiative transfer theory," *IEEE J. Sel. Areas Commun.*, vol. 26, no. 9, pp. 1620–1627, Dec. 2008.

Part III

Particle-based Propagation

6

Molecular Communication

CONTENTS

Chemical signals render flow of constituent molecules using different physical phenomena, like, diffusion and advection, through any medium; a process referred to as molecular communications (MC). Molecular diffusion is guided by Fick's laws of diffusion. Diffusion

DOI: 10.1201/9781003213017-6

is the physical process where molecular agitation and/or small-scale turbulent motions act to move the substance randomly with respect to the mean motion of the fluid. Fick's laws state that the molar flux due to diffusion is proportional to the concentration gradient. They also state that the rate of change of concentration at a point in space is proportional to the second derivative of concentration over space. The amount of substance traversing the cross-section over which the count is performed depends on the nature of the transporting process. MC environment and the interactions within it are even more complicated than wave-based communication scenarios due to the varied characteristics of the constituent molecules and the environments through which they propagate. This chapter provides a glimpse of how diffusion-based MC systems work and different modeling attempts that have been made over literature for different MC processes and different scenarios.

6.1   Introduction

The use of electromagnetic waves is neither recommended nor efficient for carrying information in certain environments, like networks of tunnels, pipelines, saltwater environments, body fluids, and blood. For nano-scale or micro-scale environments, electromagnetic communication is extremely challenging as the size of antenna required to transmit or receive depends on the wavelength of the signal. In the above-mentioned critical environments, light wave or optical communication is also not suitable as it requires either a guided medium or a direct LoS. The promising solution in such critical environments is to use chemical signals as the carrier of information [1, 2].

Chemical signals refer to the flow of constituent molecules using different physical phenomena, like diffusion and advection. through any medium, a process referred to as molecular communications (MC). The probability of MC-based signal reaching the receiver in several challenging environments is much higher than EM wave-based communications. Small particles, like molecules, lipid vesicles, are transmitted by a source into any kind of medium through which these particles propagate until they reach the receiver. The receiver, in turn, detects and decodes the information. At the source, the particles are encoded with information before they are released into the media [3, 4, 5, 6, 7].

MC signals are bio-compatible and need very little energy to generate and transmit and therefore can be used in many naturally occurring signal propagation scenarios, like chemical signals for inter- and intra-cellular communications, pheromones between insects of the same species, chemical signal used by microbes for detecting and communicating with other microorganisms. Moreover, the entire communication process requires very little energy and dissipates very low heat. An overall schematic diagram of an MC system is provided in Fig. 6.1. At the transmitter, information particles like biological or synthetic compounds, few nanometer (nm) or micrometer (μm) in size are generated stored and released. The transmitter involves a processing unit that converts the information particles to either chemical or electrical signal using energy from the environment or any other electrical source.

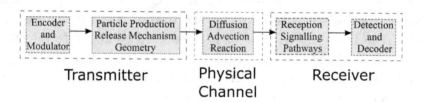

FIGURE 6.1
Overall schematic of an end-to-end MC system.

Once the information particles are released by the transmitter, they propagate through the relevant media by means of naturally occurring phenomena, like diffusion flow or assisted diffusion flow. Assisted diffusion is executed through engineered transport systems using molecular motors. On the receiver side, the signal is detected and demodulated either using a sensor, receptor, or a detector. Once detected information is estimated from the received signal using some arrival property of the information particle, like its concentration, time of arrival, or other properties. The receiver also derives its energy from the environment or any other form of power sources.

Depending on the range of communications, MC can be classified as short-range (nm scale), mid-range (μm to cm scale), or long-range (cm to m scale) communications. Short and mid-range can be used for micro-scale and long-range can be used for macro-scale communication scenario. Based on the range and scenario of communications, different media and carriers of information can be used. For example, neurotransmitters use diffusion over short range, motor proteins use active transportation over mid range and hormones over long range.

6.2 Transmitter Model

Information particles and the structure of the transmitter affect the overall flow of the signal in an MC system. Major factors on the transmitter site affecting the overall MC include geometry of the transmitter, kind of process used to generate information particles, and the release mechanism controlling the release of particles. Taking these properties into account, various modeling approaches have been proposed for molecular transmitter [8, 9]. A few worth mentioning points here are summarized below.

6.2.1 Point Transmitter

A point transmitter is modeled as a zero-dimensional point that releases A molecules instantaneously, which start flowing through the medium immediately. Therefore, the effects of the particle generation and release mechanisms are neglected. The channel impulse response (CIR) model of a point transmitter is given by $h^\bullet(t, d_0)$, where CIR is a function of time t and source position d_0.

6.2.2 Volume Transmitter

Here, the transmitter is modeled as an entity where the molecules are distributed over the transmitter volume. Information particles are generated instantaneously and are released to the medium through which they travel via diffusion. The CIR model of a volume transmitter can be expressed as

$$h(t) = \frac{1}{V_{t_x}} \int_{d \in V_{t_x}} h^\bullet(t, ||\mathbf{d} - \mathbf{d}_{r_x}||) d\mathbf{d} \tag{6.1}$$

where V_{t_x} is the volume of the transmitter, $\mathbf{d}_{r_x}$ is the distance between the transmitter and the receiver, and $\mathbf{d}$ is the distance of an arbitrary point from the transmitter along the line between the transmitter and the receiver.

6.2.3 Ion Channel-based Transmitter

Spherical objects with ion channels embedded in their membranes are used to model ion channel-based transmitters. The opening and closing of the ion channels can be controlled using a gating parameter in terms of a voltage or a ligand. So once the gating parameter is applied to the transmitter, ion channels open up releasing A molecules into the medium through which they travel via diffusion. The CIR of an ion channel-based transmitter can be given by

$$h(t) = \frac{a_{t_x}}{d_0 \sqrt{2Dt}} e^{-\frac{d_0^2 + a_{t_x}^2}{4Dt}} \sinh\left(\frac{d_0 a_{t_x}}{2Dt}\right) \tag{6.2}$$

where a_{t_x} is the radius of the information molecule and D is the diffusion coefficient of the medium given by the so-called Einstein's relation $D = kT/6\pi\eta R$, where k is the Boltzmann constant, T is the temperature in Kelvin, η is the viscosity of the medium, and R is the radius of the information particle.

6.3 Propagation Phenomenon

In MC, information particles propagate through any medium using three different physical phenomena: a) diffusion, b) advection, and c) reaction [10, 11, 12, 13, 14, 15].

6.3.1 Free Diffusion

Thermal vibration and collision between different molecules within any medium affects the flow of the particles within the medium. Resulting movement is random without moving in any particular direction and this kind of movement is referred to as Brownian motion or random walk. If the position of the ith molecule is denoted by a vector $\mathbf{d}_i(t) = [x, y, z]$, then the random walk is modeled by

$$\mathbf{d}_i(t + \Delta t) = \mathbf{d}_i(t) + \mathcal{N}(\mathbf{0}, 2D\Delta t\mathbf{I}) \tag{6.3}$$

where Δt is the time step-size, D is the diffusion coefficient of the ith molecule, $\mathcal{N}(\mu, \Sigma)$ denotes multivariate Gaussian distribution with mean μ, covariance Σ, $\mathbf{0}$ is an all-zero vector, and $\mathbf{I}$ is an identity matrix. If we denote the concentration of the solute molecules, i.e., the average number of solute molecules per unit volume by $c(\mathbf{d}, t)$, variation in $c(\mathbf{d}, t)$ can be caused by the random movement of the molecules across space and time. This variation in concentration can be calculated using Fick's second law of diffusion

$$\frac{\delta c(\mathbf{d}, t)}{\delta t} = D\nabla^2 c(\mathbf{d}, t) \tag{6.4}$$

where $\nabla^2 = \frac{\delta^2}{\delta x^2} + \frac{\delta^2}{\delta y^2} + \frac{\delta^2}{\delta z^2}$ is the Laplace operator in Cartesian coordinates. Depending on the communication scenario, (6.4) can be solved using initial and boundary conditions for that particular scenario. A glimpse of how diffusion-based MC systems work is provided in Fig. 6.2.

6.3.2 Advection

Depending on the position $\mathbf{d}$ and time $\mathbf{t}$, a velocity vector $\mathbf{v}(\mathbf{d}, t)$ can be used to characterize the process of advection. The

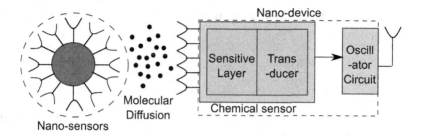

FIGURE 6.2

Physical model of a diffusion-based MC system.

movement of the ith particle from position after a time interval of $(t + \Delta t)$ due to advection can be modeled by

$$\mathbf{d}_i(t + \Delta t) = \mathbf{d}_i(t) + \mathbf{v}(\mathbf{d}_i, t)\Delta t \qquad (6.5)$$

where Δt is negligible enough to maintain constant velocity between $\mathbf{d}_i(t)$ and $\mathbf{d}_i(t + \Delta t)$. Two different kinds of advection phenomena are observable referred to as force-induced drift and bulk flow.

6.3.2.1 Force-Induced Drift

Information particles can be forced to move via advection by applying external force on them. The velocity vector for the particles, in that case, can be calculated using Stoke's law

$$\mathbf{v}(\mathbf{d}, t) = \mathbf{F}(\mathbf{d}, t)/\xi \qquad (6.6)$$

where $\mathbf{F}(\mathbf{d}, t)$ is the external force applied and ξ is the friction coefficient of the medium such that $\xi D = kT$, it is intuitive that $\mathbf{F}(\mathbf{d}, t)$ varies both with time and space.

6.3.2.2 Bulk Flow

If the movement of the particles is induced by the movement of the medium (liquid or gas) itself, the transport of molecules via advection is referred to as bulk flow. Bulk flow can be turbulent or laminar. If rough surfaces and high flow velocities contributed to the stochastic variations in the bulk flow velocity over space and time, the flow is referred to as turbulent. If the flow is not turbulent, the flow is referred to as laminar.

The change in molecular concentration caused by advection is characterized by

$$\frac{\delta c(\mathbf{d}, t)}{\delta t} = -\nabla \cdot (\mathbf{v}(\mathbf{d}, t), c(\mathbf{d}, t)) \qquad (6.7)$$

where $\nabla = \left[\frac{\delta}{\delta x}, \frac{\delta}{\delta y}, \frac{\delta}{\delta z}\right]$ is the gradient operator and $(\cdot)$ denotes the inner product of two vectors $\mathbf{x}$ and $\mathbf{y}$. Using different numerical methods, (6.7) can be solved to obtain the velocity vector.

6.3.3 Advection-Diffusion

Some applications, like drug delivery through blood capillaries, use both advection and diffusion phenomena in combination with each other. The change in molecular concentration, in this case, can be given by the following equation reflecting the combination of both processes

$$\frac{\delta c(\mathbf{d}, t)}{\delta t} = D\nabla^2 c(\mathbf{d}, t) - \nabla \cdot (\mathbf{v}(\mathbf{d}, t), c(\mathbf{d}, t)) \qquad (6.8)$$

The equation in (6.8) can be solved for velocity $\mathbf{v}(\mathbf{d}, t)$ by assuming constant uniform flow in an unbounded environment and then using the concepts of Pe'clet number and dispersion factors. Pe'clet number is used to measure the level of correlation between the diffusion and advection processes within any fluid medium. Pe'clet number can therefore be used to numerically define the relationship between advection and diffusion processes, when both are used for transporting the molecules.

6.3.4 Chemical Reaction

MC system can also employ chemical reaction for propagation of information particles. A generalized representation of chemical reaction is given by

$$\sum_{I \in \mathcal{I}} n_I I \rightarrow K \sum_{J \in \mathcal{J}} n_J J \qquad (6.9)$$

where $I \in \mathcal{I}$ are the set of reactant molecules and $J \in \mathcal{J}$ are the set of product molecules, n_I and n_j are non-negative integers acting as coefficients and K is the reaction rate constant. If the concentrations of the reactants and the product molecules are denoted by, $c_I(\mathbf{d}, t)$ and $c_J(\mathbf{d}, t)$, respectively, the rate of change of concentrations is given by

$$\frac{\delta c_I(\mathbf{d}, t)}{\delta t} = -n_I \, f(K, c_I, \forall I \in \mathcal{I})$$

$$\frac{\delta c_J(\mathbf{d}, t)}{\delta t} = -n_J \, f(K, c_J, \forall J \in \mathcal{J}) \qquad (6.10)$$

where $f(K, c_I, \forall I \in \mathcal{I})$ and $f(K, c_J, \forall J \in \mathcal{J})$ are the reaction rate functions defined by parameters like reaction rate constant and concentration of the reactant and the product molecules given by,

$$f(K, c_I, \forall I \in \mathcal{I}) = K \prod_{I \in \mathcal{I}} c_I^{\epsilon_I}(\mathbf{d}, t) \qquad (6.11)$$

where ϵ_I is the reaction order with respect to I-type reactant molecule and $\sum_{I \in \mathcal{I}}$ is the overall reaction order.

6.3.5 Modeling the Channel

Physical channel through which the particles flow between the transmitter and the receiver is affected by different factors and noise as it happens with other forms of propagation channel. The physical factors however affecting this kind of channel are quite different than EM or acoustic channels including a) constructive or disruptive advection depending on the strength and direction of the velocity vector, b) geometry of the physical channel and c) degradation and production of the molecules in order to characterize the above-mentioned factors. Several channel models have been proposed over literature. The first set of models represents bounded physical channels referred to as duct channels. Bounded channels involve transmission of molecules through the process of diffusion and the diffusion process equations can be solved by appropriate boundary conditions defined by the physical and chemical properties of the medium and the type of molecules. Two common types of duct channels are *rectangular duct channel* and *circular duct channel*.

6.3.5.1 Rectangular Duct Channel

For a rectangular duct channel with dimensions $-\infty < z < +\infty, 0 < x < l_x, 0 < y < l_y$ and fully reflective walls, the CIR can be expressed as

$$
h(t) = \frac{V_{r_x}}{l_x l_y}\left[1 + 2\sum_{n=1}^{\infty} e^{-Dn^2\pi^2 t/l_x^2}\cos(n\pi x_{r_x}/l_x)\cos(n\pi x_{t_x}/l_x)\right]
$$

$$
\times \left[1 + 2\sum_{n=1}^{\infty} e^{-Dn^2\pi^2 t/l_y^2}\cos(n\pi y_{r_x}/l_y)\cos(n\pi y_{t_x}/l_y)\right]
$$

$$
\times \left[\frac{1}{\sqrt{4D\pi t}}e^{-\frac{(z_{r_x}-z_{t_x})^2}{4Dt}}\right] \tag{6.12}
$$

where D is the diffusion coefficient of the medium, transmitter is at position, $\mathbf{d}_{t_x} = [x_{t_x}, y_{t_x}, z_{t_x}]$ receiver is at location, $\mathbf{d}_{r_x} = [x_{r_x}, y_{r_x}, z_{r_x}]$ and V_{r_x} is the volume of the receiver.

6.3.5.2 Circular Duct Channel

For a circular duct channel with dimensions $0 < \rho < a_c, 0 < \theta < 2\pi, -\infty < z < +\infty$ in cylindrical coordinates and fully reflected walls, the CIR can be expressed as

$$
h(t) = \frac{V_{r_x}e^{-\frac{(z_{r_x}-z_{t_x})^2}{4Dt}}}{2\pi a_c^2\sqrt{\pi Dt}} \times \left[1 + \sum_{n=-\infty}^{\infty}\cos(n(\phi_{r_x}-\phi_{t_x}))\sum_{\alpha}\right.
$$

$$
\left. \times e^{-Dat^2}\frac{\alpha^2 J_n(\alpha\rho_{r_x})J_n(\alpha\rho_{t_x})}{(\alpha^2 - n^2/a_c^2)J_n^2(a_c\alpha)}\right] \tag{6.13}
$$

where the transmitter is at location $\mathbf{d}_{t_x} = [\rho_{t_x}, \phi_{t_x}, z_{t_x}]$ and the receiver is at $\mathbf{d}_{r_x} = [\rho_{r_x}, \phi_{r_x}, z_{r_x}]$, $J_n(\cdot)$ is the nth order Bessel function of the first kind with the derivative $J_n'(\cdot)$ and the derivative satisfies the condition $J_n'(\alpha a_c) = 0$.

6.3.5.3 Advection Channel

For the advection process, the signal molecules experience time-invariant velocity and therefore, we can assume that $\mathbf{v}(\mathbf{d}, t) = \mathbf{v}(\mathbf{d})$ for $t > t_0$. In case of advection, two different propagation conditions are possible, one is the uniform constant advection (UCA) and the other is the laminar flow.

If the velocity vector $\mathbf{v}$ at a point $\mathbf{d}$ in space is given by $\mathbf{v}(\mathbf{d}) = [v_x, v_y, v_z]$, $\mathbf{v}$ can be decomposed into the parallel component $v_{\parallel}$ and the orthogonal one v with respect to $d_0 = d_{t_x} - d_{r_x}$. So far a point transmitter $d_{t_x} = [0, 0, -z_{t_x}]$ and a passive receiver $d_{r_x} = [0, 0, 0]$ such that, $d_0 = [0, 0, -z_{t_x}]$. In that case, $v_{\parallel} = v$ and $v = \sqrt{v_x^2 + v_y^2}$. For an unbounded channel with UCA, the CIR can be given by

$$h(t) = \frac{V_{r_x}}{(4\pi Dt)^{3/2}} e^{-\frac{(vt)^2 + (z + z_{t_x} - v_{\parallel}t)^2}{4Dt}}. \tag{6.14}$$

6.3.5.4 Laminar Flow

Laminar flow is seen mostly in circular duct like channels with a bounded environment. Let us consider a receiver with cylindrical coordinates ranging between, $a_c - l_\rho \leq \rho_{r_x} \leq a_c$, $|\phi_{r_x}| \leq l_\phi/2$, $|z_{r_x} - d_z| \leq l_z/2$, we define two different cases: a) regime where the traveling signal suffers from dispersion i.e., $\alpha_d << 1$ where α_d is the dispersion factor of the medium, b) the regime where the travelling signal travels through the flow-dominant environment, i.e., $\alpha_d >> 1$.

6.3.6 Degradation Channel

Signaling molecules are affected by degradation and production on their way to the receiver. Based on observation on the generalized behavior of MC channels, it is possible to group them into two categories: first order degradation and enzymatic degradation.

First-order Degradation This kind of degradation process refers to the transformation of one molecule to another molecule depending on the reaction rate of the medium. The CIR over the channel experiencing first-order degradati—on can be written as

$$h(t) = \tilde{h}(t)e^{-Kt} \tag{6.15}$$

where $\tilde{h}(t)$ can be given by equations for diffusion-based MC or by equations for advection-based MC and K is the reaction rate constant for the medium.

6.3.7 Enzymatic Channel

This kind of degradation is often observed in the communication channel between a passive receiver and a point transmitter. The CIR of the channel in that case for an unbounded environment can be given by

$$h(t) \approx \frac{V_{r_x}}{(4\pi Dt)^{3/2}} e^{-K_f C_E t - \frac{d_0}{4Dt}} + K_b C_{AE} t \qquad (6.16)$$

where K_f and K_b are forward and backward flow reaction constants for the medium, $C_E(\mathbf{d}, t)$ is the concentration of the information molecules, $C_{AE}(\mathbf{d}, t)$ is the concentration of the intermediate species, which remains constant over space and time. If $K_d \to \infty$ and $K_b \to 0$ then $C_{AE} = 0$ where K_d is the degradation constant of the media. If $K_f \to \infty$ and $K_b \to 0$, the CIR expression in (6.16) can be modified to obtain,

$$h(t) \approx \frac{V_{r_x}}{(4\pi Dt)^{3/2}} e^{-\frac{K_f K_d}{K_b + K_d} - C_E t - \frac{d_0}{4Dt}} \qquad (6.17)$$

where C_E is the total concentration of the final received molecules including both bound and unbound conditions.

6.4 Receiver Model

This set of models characterizes the properties of the receiver while the transmitter and the MC channel models are based on simplistic assumptions. Two groups of reception mechanisms are generally observed in MC systems; a) passive—the flow of signal molecules are not affected by the receiver, b) active—the signal molecules flow is affected by the receiver either through surface absorption and chemical reaction via receptors [16, 17, 18]. Using signaling pathways signal molecules are transformed into secondary ones for decoding information in the case of active reception. Signaling pathways are responsible for relaying information from the extracellular space to the organelles within the cytosol of natural living cells, thereby instigating response by the cell. For characterizing the receiver, transmitter is assumed to be a point transmitter located at $\mathbf{d}_{t_x}$ and releasing A signal molecules immediately with stimulation. The center of the receiver is located at $\mathbf{d}_{r_x}$ and $d_0 = ||\mathbf{d}_{t_x} - \mathbf{d}_{r_x}||$. For the propagation environment, an unbounded channel plagued by diffusion noises assumed. A few receiver models worth mentioning are summarized below (refer to Fig 6.3).

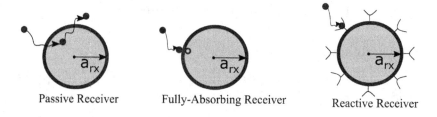

Passive Receiver   Fully-Absorbing Receiver   Reactive Receiver

FIGURE 6.3
Representation of commonly used receiver models.

6.4.1 Passive Receiver

Such receivers operate through passive reception where signal molecules arrive and leave the receiver through free diffusion. Using UCA assumption across the entire volume of the receiver, the CIR can be expressed through

$$h(t) = \frac{V_{r_x}}{(4\pi Dt)^{3/2}} e^{-\frac{d_0^2}{4Dt}} \tag{6.18}$$

where V_{r_x} is the volume of the receiver.

6.4.2 Fully Absorbing Receiver

This kind of receiver operates on active reception where the signal A molecules are absorbed via diffusion as soon as they reach the receiver. All points $\mathbf{d}$ on the receiver surface, $\mathbf{S}_{r_x}$ represents a sensing area of the receiver and the number of absorbed molecules over the small time dt represents the observed signal. The CIR in this case can be given by

$$h(t) = \frac{a_{r_x}(d_0 - a_{r_x})}{td_0\sqrt{4\pi Dt}} e^{-\frac{(d_0 - a_{r_x})^2}{4Dt}} dt \qquad (6.19)$$

and the probability of molecules being absorbed over time t can be given by,

$$g(t) = \int_{t'=0}^{t} \frac{a_{r_x}}{d_0} \operatorname{erfc}\left(\frac{d_0 - a_{r_x}}{\sqrt{4Dt}}\right) \qquad (6.20)$$

where erfc is the complementary error function.

6.4.3   Reactive Receiver

Large signaling molecules are detected by external receptors embedded in the cell membrane since they cannot passively diffuse through the cell membranes. In this case, ligand receptor complex molecules are formed through second order reaction between diffusive signaling A molecules that arrive at the cell and the receptor protein B molecules on the cell surface. The reaction can be represented by $A + B \rightleftharpoons C$, where K_f is the binding reaction rate constant and K_b is the unbinding reaction rate constant. The receiver surface covered with receptors form the sensing area of the receiver and the received signal is generated out of the number of activated receptors C.

6.5   Receive Signal Models

In this section, mathematical models are provided for signals used for system parameter estimation and data detection. Such models should be able to predict the expected number of molecules to be observed at the receiver depending on the communication scenario.

6.5.1 Deterministic Models

Assuming time invariant MC channel and independent molecule behavior, the expected number of molecules observed at the receiver at time t due to the release of N_{t_x} molecules by the transmitter at time, $\tau = 0$, can be given by $\bar{r}(t, \tau) = N_{t_x} h(t, \tau)$ where $h(t, \tau)$ is the probability of a molecule being released by the transmitter at time τ and observed at the receiver at time t, $r(t, \tau)$ is the number of molecules observed at the receiver and released by the transmitter and $\bar{r}(t, \tau) = \text{E}\{r(t, \tau)\}$ is the mean of the signal for a fixed set of channel parameters and $\text{E}\{\cdot\}$ denotes expectation.

6.5.2 Statistical Models

For time-invariant MC channels, we develop statistical models for the number of molecules observed at the receiver.

6.5.2.1 Binomial Model

Since any given molecule released by the transmitter is either observed by the receiver or not, a binary state model applies and the number of observed molecules follow the binomial distribution with N_{t_x} trials and success probability $h(t, \tau)$, $r(t, \tau) \sim \mathcal{B}(N_{t_x}, h(t, \tau))$ where $\mathcal{B}(N, p)$ represents Binomial distribution with N and p denoting the number of trials and the success probability, respectively. The probability density function (PDF) can be given by

$$f_r^{\mathcal{B}}(n) = \binom{N_{t_x}}{n} (h(t, \tau))^n (1 - h(t, \tau))^{N_{t_x} - n}$$

$$n \in \{0, 1, \dots, N_{t_x}\} \qquad (6.21)$$

Unfortunately, the binomial distribution considerably complicates the analysis of MC systems. Two approximations that can be used are:

Gaussian Model

If the expected number of molecules observed at the receiver, $\bar{r}(t, \tau)$, is sufficiently large, then we can apply the Central Limit Theorem (CLT) and approximate $r(t, \tau)$ by a Gaussian random variables with mean and variance equal to Binomial random variable (RV)

$$r(t,\tau) \sim \mathcal{N}\left(N_{t_x}h(t,\tau), N_{t_x}h(t,\tau)(1-N_{t_x}h(t,\tau))\right) \qquad (6.22)$$

with PDF

$$f_r^{\mathcal{N}}(n) = \frac{1}{\sqrt{2\pi N_{t_x}h(t,\tau)}}(1-h(t,\tau))e^{-\frac{(n-N_{t_x}h(t,\tau))^2}{2N_{t_x}h(t,\tau)}(1-h(t,\tau))} \qquad n \in \mathbb{R}$$

$$(6.23)$$

Poisson Model

For the case when the number of trials is large and the mean of the Binomial RV is small, the Binomial distribution can be well approximated by the Poisson distribution with same mean $\bar{r}(t,\tau)$, i.e., $r(t,\tau) \sim \mathcal{P}(N_{t_x}h(t,\tau))$ where $\mathcal{P}(\lambda)$ denotes Poisson distribution with λ representing the mean of the RV with probability mass function (PMF)

$$f_r^{\mathcal{P}}(n) = \frac{(N_{t_x}h(t,\tau))^n}{n!}(1-h(t,\tau))e^{-N_{t_x}h(t,\tau)} \qquad n \in \mathbb{R} \qquad (6.24)$$

6.5.3 Stochastic Model

If we consider a time-varying MC channel, the CIR can be given by

$$h(t,\tau) = \frac{V_{r_x}}{(4\pi Dt)^{3/2}}e^{-\frac{d^2(\tau)}{4D_1 t}} \qquad (6.25)$$

where $d(\tau) = \|\mathbf{d}(\tau)\|$ and $D_1 = D + D_{r_x}$ is the effective diffusion coefficient and $h(t,\tau)$ is a stochastic process with random variables $h(t,\tau_i), i \in \{1,2,3,\dots\}$. The PDF of the time variant-channel can be approximated by log-normal distribution. The above model is the most common characterization of the time-variant MC environment for performance evaluation of different MC systems.

Bibliography

[1] N. Farsad, H. Yilmaz, A. Eckford, C. Chae, and W. Guo, "A Comprehensive Survey of Recent Advancements in Molecular Communication," *IEEE Commun. Surveys Tutorials*, vol. 18, no. 3, pp. 1887–1919, third quarter 2016.

[2] L. P. Gin and I. F. Akyildiz, "Molecular Communication Options for Long Range Nanonetworks," *Computer Netw.*, vol. 53, no. 16, pp. 2753-2766, 2009.

[3] M. Pierobon and I. Akyildiz, "Noise Analysis in Ligand-Binding Reception for Molecular Communication in Nanonetworks," *IEEE Trans. Sig. Process.*, vol. 59, no. 9, pp. 4168–4182, Sep. 2011.

[4] S. Kadloor, R. Adve, and A. Eckford, "Molecular Communication Using Brownian Motion With Drift," *IEEE Trans. NanoBiosci.*, vol. 11, no. 2, pp. 89–99, Jun. 2012.

[5] T. Nakano, T. Suda, T. Koujin, T. Haraguchi, and Y. Hiraoka, *Molecular Communication Through Gap Junction Channels*. Berlin, Heidelberg: Springer Berlin Heidelberg, 2008, pp. 81–99.

[6] M. Moore, A. Enomoto, T. Nakano, R. Egashira, T. Suda, A. Kayasuga, H. Kojima, H. Sakakibara, and K. Oiwa, "A Design of a Molecular Communication System for Nanomachines Using Molecular Motors," in *Proc. Pervasive Comput. Commun. Workshops*, Mar. 2006, pp. 6 pp.–559.

[7] M. Gregori and I. F. Akyildiz, "A New Nanonetwork Architecture Using Flagellated Bacteria and Catalytic Nanomotors," *IEEE J. Select. Areas in Commun.*, vol. 28, no. 4, pp. 612–619, May 2010.

[8] I. Akyildiz, F. Brunetti, and C. Blazquez, "Nanonetworks: A New Communication Paradigm," Comput. Net., vol. 52, pp. 2260–2279, Apr. 2008.

[9] T. Nakano, M. Moore, F. Wei, A. Vasilakos, and J. Shuai, "Molecular Communication and Networking: Opportunities and Challenges," *IEEE Trans. NanoBiosci.*, vol. 11, no. 2, pp. 135–148, Jun. 2012.

[10] L. S. Meng, P. C. Yeh, K. C. Chen, and I. F. Akyildiz, "On Receiver Design for Diffusion-Based Molecular Communication," *IEEE Trans. Sig. Process.*, vol. 62, no. 22, pp. 6032–6044, Nov. 2014.

[11] M. Mahfuz, D. Makrakis, and H. Mouftah, "A Comprehensive Study of Sampling-Based Optimum Signal Detection in Concentration-Encoded Molecular Communication," *IEEE Trans. NanoBiosci.*, vol. 13, no. 3, pp. 208–222, Sep. 2014.

[12] A. Noel, K. C. Cheung, and R. Schober, "Using Dimensional Analysis to Assess Scalability and Accuracy in Molecular Communication," in *Proc. IEEE Int. Conf. Commun. (ICC)*, Jun. 2013, pp. 818–823.

[13] M. Pierobon and I. Akyildiz, "A Physical End-to-End Model for Molecular Communication in Nanonetworks," *IEEE J. Sel. Areas Commun.*, vol. 28, no. 4, pp. 602–611, May 2010.

[14] D. Kilinc and O. B. Akan, "Receiver Design for Molecular Communication," *IEEE J. Select. Areas in Commun.*, vol. 31, no. 12, pp. 705–714, Dec. 2013.

[15] W. Guo, T. Asyhari, N. Farsad, H. B. Yilmaz, B. Li, A. Eckford, and C. B. Chae, "Molecular Communications: Channel Model and Physical Layer Techniques," *IEEE Wireless Commun.*, vol. 23, no. 4, pp. 120–127, Aug. 2016.

[16] A. Noel, K. C. Cheung, and R. Schober, "Diffusive Molecular Communication with Disruptive Flows," in *Proc. IEEE Int. Conf. Commun. (ICC)*, Jun. 2014, pp. 3600–3606.

[17] V. Jamali, A. Ahmadzadeh, C. Jardin, C. Sticht, and R. Schober, "Channel Estimation for Diffusive Molecular Communications," *IEEE Trans. Commun.*, vol. 64, no. 10, pp. 4238–4252, Oct. 2016.

[18] X. Wang, M. Higgins, and M. Leeson, "Distance Estimation Schemes for Diffusion Based Molecular Communication Systems," *IEEE Commun. Lett.*, vol. 19, no. 3, pp. 399–402, Mar. 2015.

7

Quantum Field Propagation

CONTENTS

Presently the choicest mode of communication for any wireless network is the electromagnetic waves. However, today the EM spectrum is overloaded and the only available frequency bands are the ones that allow very short-range communications. The only way forward is to develop alternative modes, like quantum-domain communications. As photons are natural information carriers in fibers and free-space optical networks, their quantum nature can be exploited to carry classical information while achieving unparalleled capacity, security and reliability. This chapter starts off by explaining the analogy between the classical and the quantum world, then introducing some basic concepts related to quantum-domain communications and finally detailing the propagation phenomena related to quantum particles traversing through the classical world.

DOI: 10.1201/9781003213017-7

7.1 Introduction

The choicest mode of communications, so far, for any network is the combination of radio frequency (RF) wireless and optical fiber-based links. As new applications emerge continuously, formulating novel ways of increasing capacity of communication networks and transmission systems have become the key target. Today, the useful capacity in individual fibers has been fully exploited, requiring use of additional fibers. The situation is even worse for the wireless spectrum, where the electromagnetic spectrum is already overloaded, and the only available bands are at the frequencies that only allow very short range of communications. The only way forward is to develop alternative modes of communications [1, 2, 3, 4, 5, 6].

Quantum-domain communication is an emerging candidate for addressing challenges with traditional RF-based wireless and fiber optic-based communications. Quantum-domain communications are built on the fundamentals of the quantum physics as quantum communication deals with transfer of information using quantum particles (particles whose size is equal to or less than 10^{-9} m). Communication using electromagnetic waves, acoustic waves, magneto-inductive waves, optical waves, and molecular diffusion is referred to as classical communication as they use waves/particles larger than 10^{-9} m in size to transfer information. Classical settings cannot achieve data rates, reliability, time efficiency, and security parallel to what quantum communication can afford. The quantum world can experience phenomena, like quantum entanglement, superposition of quantum states, additivity, and superactivation of quantum channels that can be exploited to achieve unparalleled capacity, security, and reliability. An envisioned architecture for quantum-domain communication system is provided in Fig. 7.1.

As human beings, we generate classical information, and we can only understand information in classical format. To unleash the potentials of a quantum communication system in transferring

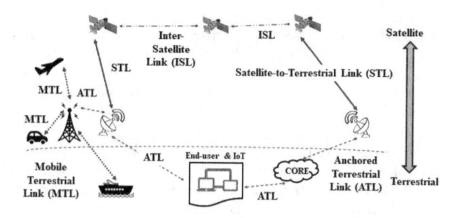

FIGURE 7.1
Unified Quantum Communications Network.

classical information, we need to encode classical information into a set of quantum states at the input of quantum channel and decode those quantum states back to classical information through measurement at the output of the quantum channel. Photons are natural carriers of information in fiber and free-space optical (FSO) networks. Since photons are a kind of quantum particles and can exhibit the fundamental phenomena of the quantum world, quantum channels can be implemented using optical fiber or FSO channels. Since this book deals with wireless communication networks, we will restrict our discussion to FSO channels and how they can be characterized when quantum particles (photons or bosons) carrying information are sent through them. Before going into detailed characterization, let us acquaint ourselves to some fundamental concepts that we use and reuse throughout this chapter.

7.2 Analogy between Classical and Quantum World

We start by relating the communication theorists' concept of channel to the physicists' description of a quantum field channel.

Classical communications can be defined as an electromagnetic field propagating in vacuum. Electromagnetic field involves EM waves that oscillate longitudinally in space pushing energy to flow in transverse direction. In order to map EM field to quantum field, we need to differentiate between the transverse and longitudinal modes of a quantum field.

The transverse mode can be characterized by its position in space in direction perpendicular to the propagation of energy and spin state or polarization. The longitudinal mode is characterized by its position in space along the direction of propagation and its position in time. Therefore a transverse mode can contain multiple longitudinal modes.

Now both for EM and quantum fields it is possible to separate orthogonal transverse mode unambiguously at the channel output; a feature always used in classical communication systems for detecting information at the receiver. Toward that end, it is possible to transmit information independently over each transverse mode at the transmitter side and therefore, each transverse mode of a quantum field can be assumed as an independent communication channel. So, when dealing with single channel capacities, we don't need to think about the transverse properties of a channel at all. We just need to distinguish between the longitudinal modes arriving at the receiver and the capacity can be calculated as the maximum number of distinguishable longitudinal modes that can arrive at the output of the channel over a certain duration of time and bandwidth operation. A simple diagrammatic representation of this correspondence between the quantum field and the classical channel is provided in Fig. 7.2.

7.3 Basic Concepts

The fundamental unit of quantum communication is a "qubit". A qubit can be defined as an abstract mathematical object that can

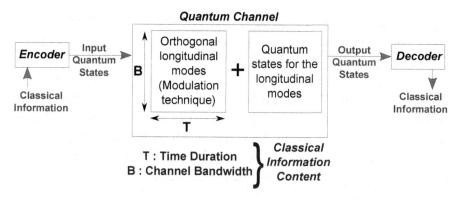

FIGURE 7.2
Correspondence between quantum field and classical channel.

have one of the two possible states $|0\rangle$ and $|1\rangle$ (corresponding to classical "0" and "1" respectively), where $|\cdot\rangle$ is referred to as the Dirac notation.

7.3.1 Qubit

A qubit can also have other possible states that are linear combinations of the two basic states $|0\rangle$ and $|1\rangle$. An arbitrary state of a qubit can be written in the format of $|\phi\rangle = \alpha|0\rangle + \beta|1\rangle$; a phenomenon known as "superposition", where α and β are complex numbers. Alternatively, the state of a qubit can be defined as a vector space, whose orthonormal basis is formed by $|0\rangle$ and $|1\rangle$ (also referred to as computational basis states).

Measurement of any qubit results in 0 with the probability of $|\alpha|^2$ and 1 with probability $|\beta|^2$ such that $|\alpha|^2 + |\beta|^2 = 1$. Physically this means that the state of a qubit is a unit vector in a two-dimensional complex vector space. For example, if a qubit is in a state, $1/\sqrt{2}|0\rangle + 1/\sqrt{2}|1\rangle$, measurement of it results in 0, 50% of the time and 1 in other 50% of time.

Therefore a qubit can exist in a continuum of states between $|0\rangle$ and $|1\rangle$ until measured. Measurement also does not preserve superposition; it collapses the state to either 0 or 1. The state $|0\rangle$

and $|1\rangle$ can be referred to as "ground" and "excited" states, respectively, when viewed with respect to the states of an electron. By focusing appropriate light of an electron, i.e., giving extra energy, it is possible to drive it from state $|0\rangle$ to $|1\rangle$ and vice versa. By controlling how long the light is focused on the electron, it is possible to move the electron into energy states between $|0\rangle$ and $|1\rangle$. It is worth mentioning here that each time an electron moves from a higher excitation state to a lower one or a ground state, packets of energy are released in the form of "photons".

The state of a qubit can also be written as

$$|\phi\rangle = \left(\cos\frac{\theta}{2}|0\rangle + e^{i\phi}\sin\frac{\theta}{2}|1\rangle \right) \qquad (7.1)$$

where θ and ϕ are real numbers. The unit three-dimensional sphere on which a point can be defined by θ and ϕ is referred to as Bloch sphere. Since there are infinite number of points on a Bloch sphere, it is possible to encode any length of information within the infinite binary expansion of θ. The state of a qubit is often represented in literature in the format $\langle 0|\phi\rangle$ which denotes the probability amplitude of state $|0\rangle$. Now states $|0\rangle$ and $|1\rangle$ are orthogonal and therefore, $\langle 0|1\rangle$ and $\langle 1|0\rangle$ are both equal to 0 and $\langle 0|0\rangle$ and $\langle 1|1\rangle$ are both equal to 1.

Let us now consider the two vectors $\frac{1}{\sqrt{2}}\binom{1}{1}$ and $\frac{1}{\sqrt{2}}\binom{1}{-1}$. They form an orthonormal basis and can be defined in terms of the computational basis as

$$|+\rangle \equiv \frac{|0\rangle + |1\rangle}{\sqrt{2}}, \quad |-\rangle \equiv \frac{|0\rangle - |1\rangle}{\sqrt{2}} \qquad (7.2)$$

$|+\rangle$ and $|-\rangle$ are referred to as Hadamard basis or diagonal basis. Now, it is also possible to calculate the amplitude of the Hadamard basis as

$$\langle +|\psi\rangle = \langle +|(\alpha|0\rangle + \beta|1\rangle) = \frac{\alpha + \beta}{\sqrt{2}} \qquad (7.3)$$

and $\langle -|\psi\rangle = \frac{\alpha-\beta}{\sqrt{2}}$. The state of a qubit can therefore be represented in terms of Hadamard basis as

$$|\psi\rangle = \left(\frac{\alpha+\beta}{\sqrt{2}}\right)|+\rangle + \left(\frac{\alpha-\beta}{\sqrt{2}}\right)|-\rangle. \qquad (7.4)$$

A qubit can be extended to a qubit system where each of the qubit can have superposition of four bases states, $|00\rangle, |01\rangle, |10\rangle$ and $|11\rangle$. In this case, the state vector of the two qubits can be written as $|\psi\rangle = a_{00}|00\rangle + a_{01}|01\rangle + a_{10}|10\rangle + a_{11}|11\rangle$. If measurement is applied to a two-qubit system, it results in $x(= 00, 01, 10, 11)$ with a probability of $|\alpha_x|^2$. Evidently $\sum_{x\in\{0,1\}^2} |\alpha_x|^2 = 1$. If after measuring the first qubit, the probability of getting the state 0 is $|\alpha_{00}|^2 + |\alpha_{01}|^2$, the post-measurement state can be expressed as

$$|\psi\rangle' = \frac{a_{00}|00\rangle + a_{01}|01\rangle}{\sqrt{|\alpha_{00}|^2 + |\alpha_{01}|^2}} \qquad (7.5)$$

The sufficient representation for a two-qubit quantum state can be given by a tensor product. If we consider two 2-dimensional vectors $\begin{bmatrix} a_1 \\ b_1 \end{bmatrix}, \begin{bmatrix} a_2 \\ b_2 \end{bmatrix}$, their tensor product can be given by, $\begin{bmatrix} a_1 \\ b_1 \end{bmatrix} \otimes \begin{bmatrix} a_1 \\ b_1 \end{bmatrix} = \begin{bmatrix} a_1 a_2 \\ a_1 b_2 \\ b_1 a_2 \\ b_1 b_2 \end{bmatrix}$. The two-qubit states can be represented also in vector format like $|00\rangle = [0; 0; 0; 0], |01\rangle = [0; 1; 0; 0], |10\rangle = [0; 0; 1; 0], |1\rangle = [0; 0; 0; 1]$, i.e., the bits inside the ket index is actually equal to 1 for the corresponding element in the vector notation.

7.3.2 Quantum Communications

Now let us consider a quantum communication scenario, where only a transmitter A is involved. A generates two qubits that are local to A itself and therefore, can be written as, $\alpha_{00}|0\rangle \otimes |0\rangle + \alpha_{01}|0\rangle \otimes |1\rangle + \alpha_{10}|1\rangle \otimes |0\rangle + \alpha_{11}|1\rangle \otimes |1\rangle = \alpha_{00}|00\rangle + \alpha_{01}|01\rangle + \alpha_{10}|10\rangle + \alpha_{11}|11\rangle$. Now let us consider a receiver B with which A shares its two-qubit state over a quantum communication channel then it can be written as, $\alpha_{00}|0\rangle^A \otimes |0\rangle^B + \alpha_{01}|0\rangle^A \otimes |1\rangle^B + \alpha_{10}|1\rangle^A \otimes |0\rangle^B + \alpha_{11}|1\rangle^A \otimes |1\rangle^B = \alpha_{00}|00\rangle^{AB} + \alpha_{01}|01\rangle^{AB} + \alpha_{10}|10\rangle^{AB} + \alpha_{11}|11\rangle^{AB}$. If this two-qubit state exhibit correlation even when shared between specially separated A and B, it can be used as a resource for communicating information between A and B process known as "entanglement."

It is possible to make two qubit states from four single qubit states, $|\phi_0\rangle$, $|\phi_1\rangle$, $|\psi_0\rangle$ and $|\psi_1\rangle$ as $|\phi_0\rangle \otimes |\psi_0\rangle = |\phi_0, \psi_0\rangle$ and $|\phi_1\rangle \otimes |\psi_1\rangle = |\phi_1, \psi_1\rangle$. Now the amplitude, $\langle\phi_1, \psi_1|\phi_0, \psi_0\rangle$ can be calculated from the dual of bra $\langle\phi_1, \psi_1|$ and ket $|\phi_0, \psi_0\rangle$. and can be expressed as $\langle\phi_1, \psi_1|\phi_0, \psi_0\rangle = \langle\phi_1, \psi_1\rangle\langle\phi_0, \psi_0\rangle$.

7.3.3   Quantum Entanglement

Let us talk about entanglement a little bit more. If A and B share an un-entangled state, then the two-qubit state can be written as $|0\rangle^A|0\rangle^B$. However, if they share an entangled quantum state, their composite state can be written as, $|\phi^+\rangle^{AB} \equiv (|0\rangle^A|0\rangle^B + |1\rangle^A|1\rangle^B)/\sqrt{2}$. This is an example of uniform superposition of the joint states $|0\rangle^A|0\rangle^B$ and $|1\rangle^A|1\rangle^B$. However it is impossible to describe is A's or B's individual states When communicated over a noiseless quantum channel i.e. we cannot directly describe the state $|\phi^+\rangle^{AB}$ as a product of the state of the form $|\phi\rangle^A|\psi\rangle^B$. This entangled state shared between A and B can be referred to as one of entanglement or one "ebit"s.

Example of a two qubit state is the Bell state or EPR (Einstein Podolski and Rosen) pair. Bell state can be written as $(|00\rangle + |11\rangle)/\sqrt{2}$. The Bell state has an interesting feature if we measure the first qubit, the second qubit measurement is predictable based on the measurement of the first qubit. For example if the first qubit is measured as 0 with probability $1/2$, the post-measurement state is given by $|\psi'\rangle = |00\rangle$. If the first qubit is measured as 1 with probability $1/2$, the post-measurement state is given by $|\psi'\rangle = |11\rangle$. This means that the measurement outcomes of the first and the second qubits of a two-qubit state are correlated. This correlation is stronger than any of its version encountered in the classical world. Four specific two-qubit states that are maximally entangled are referred to as Bell States. The resulting Bell states can be given by

$$|\phi^+\rangle^{AB} \equiv \frac{1}{\sqrt{2}} \left(|00\rangle^{AB} + |11\rangle^{AB} \right)$$

$$|\phi^-\rangle^{AB} \equiv \frac{1}{\sqrt{2}} \left(|00\rangle^{AB} - |11\rangle^{AB} \right)$$

$$|\psi^+\rangle^{AB} \equiv \frac{1}{\sqrt{2}} \left(|01\rangle^{AB} + |10\rangle^{AB} \right)$$

$$|\psi^-\rangle^{AB} \equiv \frac{1}{\sqrt{2}} \left(|01\rangle^{AB} - |10\rangle^{AB} \right) \tag{7.6}$$

These Bell states form the orthonormal basis called the Bell-basis of a two-qubit space.

7.3.4 Qudit

Another fascinating property of the quantum world is that it is possible to extinct qubit to an arbitrary d dimensional system and the corresponding quantum state is known as "qudit state." Let us consider an arbitrary qudit $|\psi\rangle = \{|k\rangle\}_{k=0}^{d-1}$ of dimension d and unknown state traveling through two noisy quantum channels $\mathcal{A}$ and $\mathcal{B}$, and let us assume that $\mathcal{A}$ is the *cyclic shift* channel and $\mathcal{B}$ is the *cyclic clock* channel. The cyclic shift channel $\mathcal{A}$ performs a cyclic permutation of the basis states and maps the set, $\{|0\rangle, |1\rangle, \dots, |d-1\rangle\}$ to $\{|1\rangle, |2\rangle, \dots, |0\rangle\}$ and vice versa with a probability p, leaving the qudit unaltered with probability $1-p$ such that

$$\mathcal{A}(|\psi\rangle) = (1-p)|\psi\rangle + pX_d|\psi\rangle \tag{7.7}$$

where X_d denotes the generalized Pauli-X gate given in Table 3.1.

The cyclic clock channel $\mathcal{B}$ introduces relative phase shift of $\omega = e^{2\pi i/d}$ between complex set of amplitudes with probability q, leaving the qudit unaltered with probability $1-q$ such that

$$\mathcal{B}(|\psi\rangle) = (1-q)|\psi\rangle + qZ_d|\psi\rangle \tag{7.8}$$

where Z_d denotes the generalized Pauli-Z gate given in Table 3.1. The complex set of amplitudes are given by α_k such that, $|\psi\rangle = \sum_{k=0}^{d-1} \alpha_k |k\rangle$, where $\alpha_k \in \mathbb{C}^d$, $\sum_{k=0}^{d-1} |\alpha_k|^2 = 1$, $\mathbb{C}^d$ is the computational basis of d-dimensional Hilbert space, $\mathcal{H}^d$. A n-qudit state, in the tensor product Hilbert space is given by $\mathcal{H}^{\otimes n} = (\mathbb{C}^d)^{\otimes n}$.

The standard basis of $\mathcal{H}^{\otimes n}$ is the orthonormal basis given by the d^n classical n-qudits such that

$$|k_1 \ldots k_n\rangle = |k_1\rangle \otimes \ldots |k_n\rangle \tag{7.9}$$

where $0 \leq k_i \leq d-1$ (for $i = 1, \ldots, n$). Since a qudit itself is an arbitrary superposition of orthonormal basis states $|k\rangle$ for d-dimensional quantum system, quantum switch for qudits will operate in a bit different way with respect to quantum switch for qubits. For example, quantum circuit architectures completely described by instances of the CNOT gate cannot implement a transposition of a pair of qudits for dimension $d > 2$. Therefore, in order to design quantum switch for qudits, we need to consider the problem of generalizing the design beyond the qubit settings.

Our model of qudit teleportation protocol consists of two processes: Alice and Bob. The sender Alice possesses a qudit of unknown state

$$|\psi\rangle_\alpha = \sum_{k=0}^{d-1} \alpha_k |k\rangle_d \tag{7.10}$$

that is to be teleported to Bob. This qudit is teleported using maximally entangled Einstein, Podolsky, and Rosen (EPR) pair of $|\Phi^+\rangle$ generated by entangling qudits $|\Phi\rangle_1$ and $|\Phi\rangle_2$ such that, $|\Phi\rangle_{12} = \frac{1}{\sqrt{d}} \sum_{k=0}^{d-1} |k\rangle_1 |k\rangle_2$, through local quantum operations and classical communications.

Firstly, CNOT Left shift operation is applied to qudits $|\psi\rangle_\alpha$ and $|\Phi^+\rangle$ followed by Hadamard operation (also defined in Table 3.1) applied to $|\psi\rangle_\alpha$. Finally, the qudits $|\psi\rangle_\alpha$ and $|\Phi^+\rangle$ are measured resulting in classical values ranging between 0 and $d-1$. Using the classical values Bob can perform necessary unitary operations to recover the original state $|\psi\rangle_\alpha$. To begin with, Alice will make a joint von Neumann measurement of the qudit $|\psi\rangle_\alpha$ and the qudit $|\Phi\rangle_1$. To make this measurement, Alice will use the Bell basis of the qudits,

$$|\Psi_{yz}\rangle = \frac{1}{\sqrt{d}} \sum_{k=0}^{d-1} e^{2\pi iyk/d}|k \oplus z\rangle|k\rangle \qquad (7.11)$$

where $k \oplus z$ means sum of k and z modulo d, $y, z = 0, 1, \ldots, d-1$ and $|\Psi_{yz}\rangle$ form a set of basis vectors for a two-qudit system which can be inverted to obtain

$$|st\rangle = \frac{1}{\sqrt{d}} \sum_{y,w=0}^{d-1} e^{-2\pi iyk/d}\delta_{t,s\oplus w}|\Phi^{yw}\rangle. \qquad (7.12)$$

The combined state of the system $|\psi\rangle_\alpha|\Phi\rangle_{12}$ in terms of the above Bell basis vectors for the system $|\psi\rangle_\alpha|\Phi\rangle_1$ is as follows:

$$|\psi\rangle_\alpha|\Phi\rangle_{12} = \frac{1}{d} \sum_{y,v=0}^{d-1} |\Psi_{yv}\rangle_{\alpha 1} U_{yv}^\dagger|\chi\rangle_2 \qquad (7.13)$$

where these unitary operators U_{vw} are given by

$$U_{vw} = \sum_{k=0}^{d-1} e^{2\pi ivk/d}|k\rangle\langle k \oplus w| \qquad (7.14)$$

and obey the following orthogonality conditions, $\mathrm{Tr}(U_{vw}^\dagger U_{yz}) = d\,\delta_{vy}\delta_{wz}$, where Tr denotes trace of a square matrix and δ_{vy} and δ_{wz} are the Kronecker delta functions.

In summary, Alice makes the von Neumann measurement in the Bell basis to obtain one of the possible d^2 results. She conveys the results of her measurement to Bob by sending $2\log_2 d$ classical bits of information. After receiving this information, Bob uses appropriate unitary operator U_{vw} on his qudit to convert its state to that of the input state, thereby completing the standard teleportation of a qudit $|\psi\rangle_\alpha$ of arbitrary state over quantum channels.

Though quantum teleportation with the aid of quantum entanglement looks promising, it is a very fragile resource and is easily degraded by noise, resulting in loss of teleported information. However, the situation can be ameliorated by exploiting quantum superposition of different causal orders realized through a quantum switch.

7.4   Propagation Phenomenon

In an FSO quantum communication channel, two different types of channels are encountered: the uplink (ground to satellite) and the downlink (satellite to ground). The uplink and the downlink quantum channels pose different challenges as the transmitter and the receiver in each case are located in a different layer of atmosphere. Different layers of atmosphere affect propagation of quantum particles differently and most turbulence and environment-related loss is experienced in the troposphere near the terrestrial ground station. Most of the propagation modeling studies on free-space quantum channels consider photon or quantum light as the mode of propagation. Therefore, in case of uplink channels, turbulence is higher near the transmitter while it is more near the receiver for the downlink channels.

In uplink channels, beam wandering is the most dominant effect since the uplink optical beam is much narrower than the large-scale turbulent eddies. Beam wandering results from the random deviation of the transmitted light beam if the beam-width is narrower than the large-scale turbulent eddies. Beam wandering results in temporal fluctuations/fading. Another impact experienced by the uplink channel is turbulence-induced beam spreading. Beam spreading is the phenomenon of beam width fluctuations occurring randomly due to atmospheric turbulence and the width fluctuation occurs in the receiver aperture plane.

In the downlink channel, photonic losses forced by diffraction are the most dominant effect. Diffraction is caused to when a light passes round the corner or through an opening whose size is smaller than the wavelength of light. The width, length, and thickness of the opening also affects the diffraction intensity and form of diffraction pattern. Another impact experienced over the down-link channel is atmospheric scintillation. Scintillation refers to the temporal variation in received irradiance and spatial variation in receiver aperture size due to the small scale turbulent eddies. An envisioned network architecture for probabilistic quantum communication network is provided in Fig. 7.3.

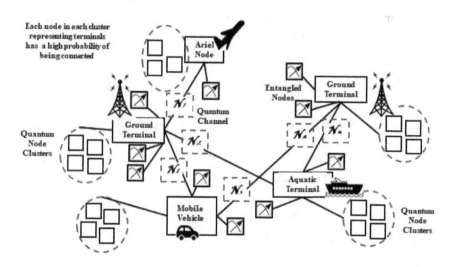

FIGURE 7.3
Probabilistic Quantum Communication Network.

7.4.1 Free-space Quantum Light

Physical phenomena mentioned, so far, like beam wandering, beam spreading, scintillation, depressions are experienced in case of both classical and quantum lights. However quantum lights also have non-classical properties, like sub-Poissonian statistics of photocounts, quadrature squeezing, non-positivity of the Glauber-Sudershan P-function and such properties are affected by the turbulence of the atmosphere. The probability distribution $p(n, T)$ of the number of counts n from a photoelectric detector illuminated by coherent light for a time T approaches the Poisson distribution for classical particles (electrons) at low degeneracies and the distribution characteristic of classical waves (electromagnetic waves) at high degeneracies. At any point in time t, distribution of n can be given by

$$p(n, t, T) = \frac{1}{n!} \left[\alpha \int_t^{t+T} P(t')dt' \right]^n e^{-\alpha \int_t^{t+T} P(t')dt'} \qquad (7.15)$$

where α is a constant representing quantum sensitivity of the photodetector and $P(t)$ represents the classical intensity of the quantum light beam. Equation (7.15) can be written as

$$p(n, T) = \frac{(\bar{n}\xi/T)^n}{(1 + \bar{n}\xi/T)^{n+T/\xi}(T/\xi - 1)\mathrm{B}(n+1, T/\xi - 1)} \qquad (7.16)$$

where T/ξ cells make up the phase space where $\mathcal{S} = T/\xi$, and

$$p(n, \mathcal{S}) = \frac{(\bar{n}\mathcal{S})^n}{(1 + \bar{n}\mathcal{S})^{n+\mathcal{S}}(\mathcal{S} - 1)\mathrm{B}(n+1, \mathcal{S} - 1)} \qquad (7.17)$$

where $\bar{n}$ is the average photon count over time and phase space and $\mathrm{B}(\cdot, \cdot)$ is the beta function given by $\mathrm{B}(x, y) = \int_0^1 t^{x-1}(1 - t)^{y-1}dt$ for complex inputs x and y such that $Re\{x, y\} > 0$. In (7.17), the degeneracy parameter $\bar{n}\mathcal{S}$ plays an important role in guiding the behavior of the distribution in (7.17), when the degeneracy is very small $\mathcal{S} \gg \bar{n}$ then, (7.17) reduces to the classical Poisson distribution

$$p(n, T) \approx \frac{e^{-\bar{n}\mathcal{S}^n}}{(\bar{n}\mathcal{S})^n n!} = \frac{\bar{n}^n}{n!} e^{-\bar{n}}. \qquad (7.18)$$

However, if the light is so weak (i.e. $\frac{\bar{n}}{\mathcal{S}} \ll 1$), the degeneracy is unlikely to be detected in measurement of a single beam of quantum light on a single boson carrying information. This situation is referred to as sub-Poissonian light and sub-Poissonian in quantum light will be even more severely affected by the atmospheric turbulence. When the degeneracy is high, $p(n, T)$ exhibits a little bit different behavior referred to as super-Poissonian light. If

$n, \bar{n} >> \mathcal{S}$, we can expand $n!$ and $\Gamma(n + \mathcal{S})$ by Stirling's theorem and (7.18) can be modified to obtain

$$p(n, T) \approx \left(\frac{\mathcal{S}}{\bar{n}}\right)^{\mathcal{S}} \frac{e^{-n\mathcal{S}/\bar{n}} n}{\Gamma(\mathcal{S})} \tag{7.19}$$

where (7.19) is the guiding equation for Super-Poissonian light [7, 8, 9, 10, 11].

In physics, a squeezed state is a quantum state that is usually described by two non-committing observables having continuous Spectra of eigenvalues. Physically, this can be interpreted as the circle denoting the uncertainty of a coherent state in the quadrature phase space has, that is being squeezed to an ellipse of the same area. The basic idea behind squeezing can be explained using quantum harmonic oscillator, whose wave-function in vacuum can be given by

$$\psi_0(x) = \frac{1}{4\sqrt{\pi}} e^{-x^2/2} \tag{7.20}$$

with momentum basis of

$$\tilde{\psi}_0(p) = \frac{1}{4\sqrt{\pi}} e^{-p^2/2} \tag{7.21}$$

with variance in vacuum state as $\langle 0|\sigma_x^2|0\rangle = \langle 0|\sigma_p^2|0\rangle = 1/2$ where σ_x^2 and σ_p^2 are the variances of the position and the momentum observables. For the squeezed-vacuum state, the position and momentum operators change to

$$\psi_R(x) = \frac{\sqrt{R}}{4\sqrt{\pi}} e^{-(Rx)^2/2} \tag{7.22}$$

and

$$\psi_R(p) = \frac{1}{\sqrt{R}4\sqrt{\pi}} e^{-(p/R)^2/2} \tag{7.23}$$

where $R > 0$ is the squeezing factor and the corresponding variances are given by, $\sigma_x^2 = 1/2R^2$ and $\sigma_p^2 = R^2/2$. If $R > 1$; it is referred to as position-squeezed and if $R < 1$ the state is referred to as momentum-squeezed. If the quantum field quadrature is given by, $\tilde{x}_\theta = \tilde{x}\cos\theta + \tilde{p}\sin\theta$, squeezing in any of the quadrature is referred to as the quadrature-squeezing.

Squeezing can be easily viewed using the Wingar function. Generating squeezed estates for transmitting quantum light is extremely crucial as in squeezed states, it is possible to experience noise lesser than the condition where there is no light at all (i.e., vacuum) squeezed states, in practicality, suffer from losses present in the transmitter, channel, and the receiver. The losses over the entire communication system affecting the squeezed states can be encompassed within the transmissivity coefficient T. If we consider a transmitter with signal mode $\hat{a}$ and there exists our dimensionless $\hat{U}$ mode, the interference between them can then be expressed as, $\hat{a}' = \tau\hat{a} - \rho\hat{v}$. Here, $\tau^2 = T$ and $\rho^2 = 1 - T$, the transmissivity and reflectivity of the communication system using quantum light for carrying information, respectively.

7.4.2   Atmospheric Quantum Channel

We consider a communication model depicted in Fig. 7.1, where a light source access a transmitter and a telescope is used as a receiver and the transmission of information occurs over the atmosphere. We assume that a balanced homodyne detector is used at the receiver side for filtering out unwanted modes of quantum light and the background radiation through the use of a local oscillator. The local oscillator uses the principle that the difference of photocurrents at the detector is proportional to the field quadrature of the required output mode. Complete information on the quantum state of the output mode can be obtained from the knowledge on the quadrature distributions of the detected quantum field. Finally, homodyne detection is used to reconstruct the associated Wingar function, photon-number distribution, movement of the radiation field, and the density matrix of the operator [12, 13, 14, 15, 16].

A linear attenuating process is used to characterize the propagation of the quantum light through the atmosphere. Different shapes of the local oscillator pulses at the homodyne detector results in signal losses owing to absorption, scattering, and mode-mismatch of the quantum light. In such a scenario, the radiation between the input mode, $P_{in}(\alpha)$ and the attenuated output mode $P_T(\alpha)$ can be expressed using a Glauber-Sudarshan P function as

$$P_T(\alpha) = \frac{1}{|T|^2} P_{in}\left(\frac{\alpha}{T}\right) \quad (7.24)$$

where T is the transmissivity of the medium. The linear attenuation model can be modified to model the turbulent atmosphere by representing T as a random variable with varying magnitude and phase. Consequently the probability distribution of T, $p(T)$ can be used to average $P_T(\alpha)$ and the P function of the output mode $P_{out}(\alpha)$ can in turn be expressed as

$$P_{out}(\alpha) = \int_{|T|^2 \leq 1} d^2 T p(T) \frac{1}{|T|^2} P_{in}\left(\frac{\alpha}{T}\right) \quad (7.25)$$

where $\int_{|T|^2 \leq 1}$ represents integration of a circular area. The main challenge in solving (7.25) is to arrive at a suitable distribution for T that is a mathematically tractable expression for $p(T)$. There are different methods of formulating $p(T)$. In the first method, let us consider k discrete turbulent eddies each with a random transmissivity T_k. Eddies are swirls of fluids and are created due to reverse current experienced in turbulent flow regime and therefore we can write, $T = \prod_k T_k$. If $t = |T|$ represents the magnitude and $\phi = \arg(T)$ represents the phase, $p(T)$ is a two-dimensional distribution with lognormal distributed t and normally distributed ϕ and we can write

$$p(t, \phi) \approx \frac{1}{2\pi t \alpha_\theta \alpha_\phi \sqrt{1 - S^2}} e^{-\frac{1}{2(1 - S^2)}\left[\left(\frac{\ln t + \bar{\theta}}{\alpha_\theta}\right)^2 + \left(\frac{\phi}{\alpha_\phi}\right)^2 + 2S \frac{\ln t + \bar{\theta}}{\alpha_\theta} \frac{\phi}{\alpha_\phi}\right]}$$

$$(7.26)$$

where $\theta = -\ln t$, $\bar{\theta}$ and α_θ are the mean and variance of θ, respectively, α_ϕ is the variance of ϕ and S is the correlation coefficient between θ and ϕ.

The second method relies on experimental and simulation-based results. In this case, let us assume that the input field exists in a coherent state $|\gamma\rangle$. In this approach, the characteristic function $\phi_{out}(\beta)$ of the P-function of the output state is used to calculate the distribution $p(T)$. The transmissivity T is assumed to be complex with real part T_r and the imaginary part T_i. Therefore we can write

$$p(T_r, T_i) = \frac{1}{4} \sum_{n,m=-\infty}^{+\infty} \phi_{out}\left(\frac{\pi}{2\gamma^*}[m+in]\right) e^{i\pi(mT_i - nT_r)}. \qquad (7.27)$$

Since we are considering homodyne receiver with fixed local oscillator amplitude r, phase $(\pi/2 - \arg\beta)$, N-size photo-counting sample and Δn_j is the difference between the photo counting events, $\phi_{out}(\beta)$ can be expressed as

$$\phi_{out}(\beta) = e^{\frac{|\beta|^2}{2}} \frac{1}{N} \sum_{j=1}^{N} e^{i\frac{|\beta|\Delta n_j}{r}}. \qquad (7.28)$$

The third method is to model $p(T)$ using its statistical moments. The moments can be measured by a balanced or homodyne detector using

$$M_{nm} = \mathrm{Tr}[\hat{\rho}\hat{a}^{\dagger n}\hat{a}^{m}] = \int_{-\infty}^{\infty} \mathrm{d}^2\alpha\, p(\alpha)\alpha^{*n}\alpha^{m} \qquad (7.29)$$

where M_{nm} is the matrix of moments of photon correlation operator for the transmitted quantum field, $\hat{a}$ is the photon creation operator of the quantum light source in absence of any dephasing effect, and $\hat{\rho}$ is the density metrics operator. Using (7.29), the relation between the moments of the input and the output modes can be formulated using the input output relationship of the quantum field. Once the properties of the input mode are known, the properties of the output mode can be obtained by using the formulation of $p(T)$.

Once $p(T)$ is formulated, we can turn our attention to finalize the expression for the P function of the output mode. It has been observed that if the P function is not positive-definite, we arrive at the non-classical state of the quantum light. Also it has been observed $p(T)$ has a strong maximum at $T = T_0$, if the atmosphere is weakly turbulent. Using these observations (7.25) can be modified using the first order Laplace approximation as

$$P_{out}(\alpha) = \int_{-\infty}^{+\infty} d^2\beta \frac{1}{|T_0|^2} P_{in}\left(\frac{\beta - \gamma}{T_0}\right) \times \frac{1}{|\gamma|^2} p\left(\frac{\alpha - \beta}{\gamma}\right) \quad (7.30)$$

where $\gamma = \langle \hat{a} \rangle$ is the displacement parameter of the input quantum field.

Next we extend our concept of free-space quantum channel model using distribution of transmission coefficient and P-functions of the input and the output modes of the quantum source and detector to the case of beam wandering and unstable adjustment of radiation source. Beam-wandering losses result from weak absorption and the aperture truncation at the receiver. Superposition of Gaussian beams with different wavenumbers k when moving along the Z-axis toward the aperture plane can experience deflection by a distance r from the center of the receiver aperture of radius a. The radius of the beams spot is denoted by W. In such a scenario, the transmission efficiency is dependent on the beam-deflection distance and the beam-spot radius.

Transmission efficiency is calculated by squaring the transmission coefficient T. Any Gaussian beam will have a transmission efficiency of $T^2(k) = \int_A |U(x, y, z_{ap}; k)|^2 dx dy$ with A the area of the aperture opening, z_{ap} is the distance along the Z-axis on the aperture plane from the quantum light source and $U(x, y, z_{ap}; k)$ is the Gaussian beam along the XY plane. If we expand $U(x, y, z_{ap}; k)$ in terms of the normalized Gaussian distribution, $T^2(k)$ can be written as

$$T^2 = \frac{2}{\pi W^2} e^{-2\frac{r^2}{W^2}} \int_0^a \rho e^{-2\frac{\rho^2}{W^2}} I_0(4/W^2 r\rho) d\rho \quad (7.31)$$

where a is the aperture radius and $I_n(\cdot)$ is the modified Bessel function. Using the maximum transmissivity, $T = T_0$, it is possible to arrive at the numerical solution for the integral in (7.31) to obtain $T^2 = T_0^2 e^{-\left(\frac{r}{R}\right)^\lambda}$ where λ and R are the shape and scale parameters, respectively, and can be calculated using the following expressions

$$\lambda = \frac{8a^2}{W^2} \frac{e^{4a^2/W^2} I_1(4a^2/W^2)}{1 - e^{4a^2/W^2} I_0(4a^2/W^2)} \left[\ln\left(\frac{2T_0^2}{1 - e^{4a^2/W^2} I_0(4a^2/W^2)}\right) \right]^{-1} \tag{7.32}$$

and

$$R = a \left[\ln\left(\frac{2T_0^2}{1 - e^{4a^2/W^2} I_0(4a^2/W^2)}\right) \right]^{-1/\lambda} \tag{7.33}$$

If we assume that the beam-center position is normally distributed with variance σ^2 around the point at distance d from the receiver aperture center, the beam deflection distance r varies according to the Rice distribution with parameters d and σ. In that case, $p(T)$ will be log-negative generalized Rice distributed given by,

$$p(T) = \frac{2R^2}{\sigma^2 \lambda T} \left(2 \ln \frac{T_0}{T}\right)^{2/\lambda - 1} I_0\left(\frac{2R^2}{\sigma^2} \left(2 \ln \frac{T_0}{T}\right)^{1/\lambda}\right)$$

$$\times e^{-\frac{1}{2\sigma^2}\left(R^2(2\ln\frac{T_0}{T})^{2/\lambda} + d^2\right)} \tag{7.34}$$

If $d = 0$, (7.34) reduces to the log-negative Weibull distribution as,

$$p(T) = \frac{2R^2}{\sigma^2 \lambda T} \left(2 \ln \frac{T_0}{T}\right)^{2/\lambda - 1} I_0$$

$$\times \left(\frac{2R^2}{\sigma^2} \left(2 \ln \frac{T_0}{T}\right)^{1/\lambda}\right) e^{-\frac{1}{2\sigma^2}\left(R^2(2\ln\frac{T_0}{T})^{2/\lambda} + d^2\right)} \tag{7.35}$$

In summary, the aperture transmission coefficient is generally log-negative generalized Rice distributed. However, if the centers of beam-wandering and aperture coincide, the aperture transmission coefficient exhibits log-negative Weibull distribution.

Next, let's extend our beam wandering free-space quantum channel model to include beam spreading and deformation. Specifically, we will shift our assumption of circular quantum light beam to an elliptical beam approximately. Random fluctuations in the refractive index of air results from the spatio-temporal variation of temperature and pressure in the atmosphere. In this case, the beam of transmitted photons (quantum light) suffers from effects like beam wandering, beam spreading, and beam deformation. If we assume a Gaussian input beam incident at a distance L from the receiver aperture, the transmittance intensity of the transmitting signal varies according to

$$\eta = \int_A d_{\mathbf{r}}^2 I(\mathbf{r}; L) \tag{7.36}$$

where A is the aperture area, $\mathbf{r}$ is the full aperture plane, and $I(\mathbf{r}; L)$ is the normalized intensity. Effects like beam wandering and beam spreading result in a deformed beam profile, which will be elliptical in shape. For elliptical beam $I(r; L)$ can be written as

$$I(\mathbf{r}; L) = \frac{2}{\pi\sqrt{\det(\mathbf{s})}} e^{-2(\mathbf{r}-\mathbf{r_0})\mathbf{S}^{-1}(\mathbf{r}-\mathbf{r_0})} \tag{7.37}$$

where $\mathbf{r_0}$ is the centroid of the elliptical beam and $\mathbf{S}$ is the real symmetric positive definite spot-shape matrix.

Beam wandering and beam spreading are predominant effects in the uplink channel. The downlink channel is much less prone to losses as the receiver antenna or the terrestrial telescope is encountered very quickly by the incident photon beam right after hitting the turbulent troposphere. Over the downlink channel, diffraction and scintillation are the most common plugging factors. Scintillation results from the variation in receiver aperture size and received beam intensity or irradiance. On the other hand, atmospheric diffraction is caused by the movement of quantum light through fine layers of dust particles trapped in the turbulent layers of the troposphere. The effects on the downlink channel can be modeled through a scaling factor, ϵ/Q where Q is the diameter of the receiver aperture and ϵ is the transmission wavelength.

Since atmospheric fluctuations occur at a rate way slower than the actual transmit data rate, it is possible to measure the channel's transmission coefficient. Several measurement campaigns or techniques have been introduced in literature for this purpose. These include: i) sending coherent or classical light pulses carrying quantum information, ii) sending a local oscillator signal through the channel mode whose polarization is orthogonal to the signal, iii) sending auxiliary classical laser beam, and iv) practically implementing a quantum communication link over the turbulent atmosphere.

Bibliography

[1] R. Ursin et al., "Entanglement-based quantum communication over 144 km", *Nature Phys.*, vol. 3, pp. 481, 2007.

[2] T. Scheidl et al., "Feasibility of 300km quantum key distribution with entangled states", *New J, Phys.*, vol. 11, pp. 085002, 2009.

[3] A. Fedrizzi, R. Ursin, T. Herbst, M. Nespoli, R. Prevedel, T. Scheidl, F. Tiefenbacher, T. Jennewein, and A. Zeilinger, "High-fidelity transmission of entanglement over a high-loss free-space channel", *Nature Phys.*, vol. 5, pp. 389, 2009.

[4] I. Capraro, A. Tomaello, A. Dall'Arche, F. Gerlin, R. Ursin, G. Vallone, and P. Villoresi, "Impact of Turbulence in Long Range Quantum and Classical Communications", *Phys. Rev. Lett.* vol. 109, 200502, 2012.

[5] J. Yin et al., "Quantum teleportation and entanglement distribution over 100-kilometre free-space channels", *Nature*, vol. 488, pp. 185, 2012.

[6] X. Ma et al., "Quantum teleportation over 143 kilometres using active feed-forward", *Nature*, vol. 489, pp. 269, 2012.

[7] X. Zhu and J. M. Kahn, "Free-space optical communication through atmospheric turbulence channels," *IEEE Trans. on Communications*, vol. 50, no. 8, pp. 1293–1300, Aug. 2002.

[8] J. H. Churnside and S. F. Clifford, "Log-normal Rician probability density function of optical scintillations in the turbulent atmosphere," *J. Opt. Soc. Am. A*, vol. 4, no. 10, pp. 1923–1930, Oct. 1987.

219

[9] R. S. Kennedy, "Communication through optical scattering channels: an introduction," *Proc. IEEE*, vol. 58, no. 10, pp. 1651–1665, Oct. 1970.

[10] J. W. Strohbehn, "Modern theories in the propagation of optical waves in a turbulent medium," in *Laser Beam Propagation in the Atmosphere*, J.W. Strohbehn ed. Springer, New York, 1978, pp. 45106.

[11] E. Jakerman, "On the statistics of K-distributed noise," *J. Phys. A*, vol. 13, no. 1, pp. 31–48, Jan. 1980.

[12] P. Milonni, J. Carter, Ch. Peterson, and R. Hughes, "Effects of Propagation through Atmospheric Turbulence on Photon Statistics", *J. Opt. B*, vol. 6, no. S742, 2004.

[13] C. Paterson, "Atmospheric Turbulence and Orbital Angular Momentum of Single Photons for Optical Communication", *Phys. Rev. Lett.*, vol. 94, pp. 153901, 2005.

[14] A. A. Semenov and W. Vogel, "Quantum Light in the Turbulent Atmosphere", *Phys. Rev. A*, vol. 80, pp. 021802(R), 2009.

[15] A. A. Semenov and W. Vogel, "Entanglement Transfer through the Turbulent Atmosphere," *Phys. Rev. A*, vol. 81, pp. 023835, 2010.

[16] D. Yu. Vasylyev, A. A. Semenov, and W. Vogel, "Toward Global Quantum Communication: Beam Wandering Preserves Nonclassicality", *Phys. Rev. Lett.* vol. 108, pp. 220501, 2012.

Part IV

Conclusion

8

The Future

CONTENTS

Till date, wireless communications has enabled billions of people living in different parts of the world to be connected to each other and exchange information. Wireless systems will continue to permeate every walk of life delivering data at any point, anywhere and anytime in the universe through integration of different communication networks (terrestrial, underwater, satellite quantum etc.). Though the wireless world is a diverse, intricate space which makes integration of different kinds of wireless networks extremely challenging, multi-mode wireless communication systems will form the backbone of future applications like connecting with exoplanets, emergency response and prediction, quantum satellite networks etc. They are becoming even more important due to the associated need for reliability in operation, link maintenance and seamless hand-off between networks. Therefore, a unified framework for modeling flow of information will lay the foundation for a completely new generation of mixed-mode communication systems capable of truly ubiquitous access to data (irrespective of position on the earth or in space), ensuring ultra-reliability and reducing latency. This chapter provides examples of some possible futuristic multi-media, mixed-mode communication systems. It also provides some probable future research directions for characterizing information flow through multi-media, mixed-mode communication systems.

DOI: 10.1201/9781003213017-8

8.1 Technological Trends

We are developing systems like distributed artificial intelligence (AI)-aided activity management for smart health applications considering my audiences are early-stage engineering students. The idea is to help diabetic patients with real-time management of their blood sugar levels with alerts and prompts on their smartphones or hand-held monitors about what kind, level, and duration of activities or exercises the patient should follow under different conditions and when to execute them.

Let's start with some basic concepts and components of a wireless communication system. The three most important components of a communication system are the signal, medium, and information and for a wireless communication system information is transported over air using electromagnetic signals. For our small system, blood glucose levels of the monitored individual is the information. Signal is generated through a combination of electrical current and magnetic field propagating in directions transverse to each other. Imagine connecting a conductor to a battery or any other energy source, some of the electrons are loosely bound to their atoms and the extra bit of energy provided by the battery helps in kicking off an electron from its own atom to fly off to the orbit of the neighboring atom, which in turn kicks of another electron in the present atom to the next one.

Now, while the electron is moving out, a hole is created within the atom temporarily, which is positively charged. So for the flow of negative charges, there is a flow of positive charges in equal and opposite direction thereby resulting in a positive electric current flow. Now, electrical current is always associated with magnetic field. Remember the right-hand thumb rule, if the current is flowing in the direction of the right-hand thumb, the magnetic field will flow in the direction perpendicular to the current along the other four fingers. So basically we have two fishes, one moving up and down and another from left to right and the resultant field is the electromagnetic signal that propagates forward. Now the question is how the electromagnetic signal can be used to carry the information on the blood glucose level.

Consider a monitored individual with a wearable patch for sensing the sugar level and sending the observation on the smartphone or a handheld meter if the glucose level is in the normal range that is 90–140 mg per liter, the sensor sends a 0 and if it is more than 140, it sends a 1. In this way, a pulse train of zeros and ones is generated by the sensor. At the handheld device, the pulse train is multiplied by a fixed frequency sinusoidal waveform or electrical signal in a way that the sinusoid changes phase as soon as there is a transition from 0 to 1 or from 1 to 0, a technique referred to as binary phase shift key or BPSK modulation. The modulated signal is an electrical signal that drives current through the antenna elements, an associated magnetic field and a resultant electromagnetic signal through the air. Now, instead of just 2 levels, we can divide the glucose range from 90 to 200 mg/L into 4, 8 or 2^M levels, where M is the number of bits used to represent each level the corresponding modulation will then be an M-ary PSK. But, why are you talking about this, it's because the more the number of levels introduced to the data, the more precise will be the output of our AI-aided system regarding the appropriate actions we need to adopt.

Now let's consider a smart system consisting of three monitored patients who sent the blood sugar readings to a gateway or a communication hub. Now we want to develop an AI-aided system. AI mimics human intelligence by acquiring knowledge from its own experience with respect to a particular task, data or environment and relies on a computer program or algorithm that helps in learning; a technique known as machine learning. Over 60 different types of machine learning algorithms exist and today I will talk about only one such algorithm called the artificial neural network or ANN.

ANN mimics the functioning of a human brain with several interconnected neurons in some pattern to allow communication between them. Neurons can be imagined as units that hold a number. ANN consists of three layers, input layer with neurons representing the glucose level data received from the patient, hidden layer with activation functions for computing relevant outputs depending on the input and the output layer with neurons representing different outputs in our case; they are the time and the duration of recommended activity depending on the patients sugar levels. The activation function can be any combination of different mathematical operations. In the first iteration, the output recommended activity is computed by the gateway using a suitable ANN model. This first iteration can be thought as an initial step when the patient is setting up his own glucose meter.

In the second iteration, the outputs and the training model are sent back to the patients' devices. The devices, in turn, train themselves and display the recommendation or prompt to the patient. The idea behind sharing the ANN model with a user device is to make the devices intelligent themselves over a certain number of iterations. Remember that sharing the model does not mean sharing user data with other users. That will be complete encroachment of user privacy. It's only the activation function or weights and biases that are shared.

In the next iteration, a new set of inputs are generated by the patients' sensors. Each patient device uses our own input to train that existing model while all the inputs are used to modify the ANN model at the hub. The modified model is again sent back to the patient and the process continues to obtain real-time output depending on the blood glucose levels of the individuals. In this way, each patient's own handle device can arrive at optimized decision on the recommended activities through distributed collaborative learning.

AI is great but the problem is it's not energy sustainable. It has been found that for training a standard model with multiple layers requires more than 600,000 pounds equivalent of carbon dioxide, which is 5 times the lifetime emission of a standard American car. Moreover, with so many applications coming up, internet of vehicles, internet of energy, internet of everything, we are going to run out of resources like computational power and memory, energy, or the electromagnetic spectrum.

Another problem rising from the financial angle of network service providers is the Digital Divide in countries like Ireland with all the population concentrated in cities and quite sparse in rural areas. So there are a lot of base stations deployed in the urban areas and a very few in the rural areas. We need a lot of signal power to overcome the interference and noise in densely deployed urban scenarios and to overcome the propagation path loss owing to the distance between the users and sparsely deployed rural base stations. So, though it may not seem likely we are unknowingly contributing to the overall carbon footprint. One solution to salvage the situation is to switch the clean renewable sources of energy from fossil fuels. However, they are budding systems and will take a few years to come up to their full potential. So the big question is what we can do as an electronic engineer. Can we make our smart environments energy efficient too?

8.2   Generalized Model

Intra-body communication promises to revolutionize the future healthcare through facilitating prediagnosis of impending medical conditions. Some of the major evolving application areas of intra-body communications include drug delivery, detection of toxic substances, micro-bacteria, viruses or allergens, early detection of cancer cells, etc.[1]. Due to the underlying challenges of using conventional Electromagnetic communications within the intra-body environment, Body-Centric Nano Networks (BCNNs)/Internet of Bio Nano things (IoBNT) have taken inspirations from the biological communication processes [2]. In this regard, the fields of

Molecular communication (MC) has evolved, which deals with exchange of molecules by the transmitter and receiver. Also, the Tera-hertz (THz) band (0.1 - 10 THz) has been regarded as one of the best candidates for intra-body communication as it offers high reliability and negligible latency [3]. The use of hybrid scheme that combines molecular and THz communication for intra-body applications has recently been proposed [2] for the timely information transmission about the ongoing physiological processes.

8.2.1 Need for a Generalized Model

In MC, transmitters, receivers, and/or actuators are deployed in the human body using a predefined application specific topology and molecules are used as information carriers. The transmitters may encode the message as number, type or release time/pattern of the molecules; this message is subsequently sensed and decoded by the receiver nodes [4]. The THz communication for intra-body applications is encouraged due to offering high bandwidth and low latency. Due to the higher speed and low latency of THz communication as compared to the MC, it is often preferred to be used over long distances within the intra-body. Also, the information from the intra-body network is transmitted to the on-body nano-devices using THz link, which is subsequently forwarded to a cloud-computing platform via an IoT Gateway, which further connects to the experts or emergency service providers for the appropriate action to be taken.

As with other biological processes which occur at microscopic scale inside the human body, it has been a serious challenge to monitor the ovulation. Ovulation refers to the process when egg/s release from the female ovaries into the Fallopian tubes and begin travelling to the uterus. In general, the eggs remain in Fallopian tubes for up to 24 hours where they can be fused with sperm. At this time, for the natural conception to occur, it is crucial that the sexual intercourse is timed near ovulation so that the male sperm could swim up to the Fallopian tube and attach itself with one of the eggs. The duo then attaches to the endometrium (uterine lining) and proceeds to the later stages of pregnancy. On the other hand, in case the egg could not fertilize when in the Fallopian tube, the female body sheds it along with the endometrium via menstrual blood.

At present, the females trying to conceive have to rely both on estimates and ultrasound scans to monitor their fertility. Ovulation approximately takes place after 13–15 days before the start of each menstrual period. When in ovaries, the eggs remain in sacs referred as follicles. The presence and size of follicles in ovaries is detected using conventional pelvic and/or trans-vaginal ultrasound scans [5]. As a follicle reaches approximate size of 18–28 mm, it is considered ready for ovulation. When the eggs are released into the Fallopian tubes, the available non-invasive tracking methods (such as measuring Luteininzing Hormone, and maintaining Basal Body Temperature Chart) provide no guarantee to detect the presence of egg in real time [6]. As a result, the probability of missing the ovulation increases even after going through a series of ultrasounds and diagnostic tests (urine and blood for LH monitoring), which in turn increases probability of failed attempts of natural conception or Intrauterine insemination (IUI) and In-vitro fertilization (IVF) treatments. Therefore, it has been proposed to use intra-body hybrid communication architecture integrated with nano-sensors/nanodevices for detecting presence of eggs in Fallopian tube in the real-time. Here, hybrid communication refers to the combined use of MC and THz communication to realize the advantages of both paradigms.

If we want to develop a system that uses intra-body communication techniques for monitoring the human fertility, we need to deploy an end-to-end architecture that integrates intra-body molecular and THz communication with body area network (BAN) and extra-body IoT backhaul network. Nano-sensors will be implanted within the Fallopian tubes that transmit alert signals to the patient and physicians as soon as an egg is released. The overall architecture can therefore be grouped into three communication modules operating over different media and modes.

- Communication Module 1: The nano-network comprising of the intra-body nano-sensors/nano-nodes and nano-transmitters/

actuators deployed within the Fallopian Tube, operating using assisted diffusion-based molecular communication.

- Communication Module 2: The nano-network between the intra-body nano-transmitters/actuators and wearable or on-body device, communicating using THz band

- Communication Module 3: The conventional IoT backhaul network for the data transmission between the wearable/on-body device to the remote sink and/or handheld device using 4G/5G or WiFi.

Therefore, we need to develop a unified framework for modeling flow of information that will lay the foundation for a completely new generation of mixed-mode communications systems which enable truly ubiquitous access to data (irrespective of position on the earth or in space), ensure ultra-reliability, and reduce latency. An envisioned architecture for the mixed-mode communication system is provided in Fig. 8.1.

8.2.2 Directions

The best way forward will be to propound a transformative way of modeling information flow wirelessly and develop a mathematical framework capable of representing the information flow over any medium; and a corresponding simulator capable of estimating the characteristics of the transmitted signal at any point in space, irrespective of geographical location and resources used. Concepts of Whitney triangulation can be used for embedding a graph onto a space surrounding the propagating signal between the transmitter and the receiver, concepts of classical harmonic oscillators to represent each of the triangles and wave equations to map flow of information and energy through the graph. Strength of the propagating field, energy delivered by the transmitter and energy dissipated due to interaction between the travelling signal and the oscillators characterizing the surrounding space also needs to be quantified.

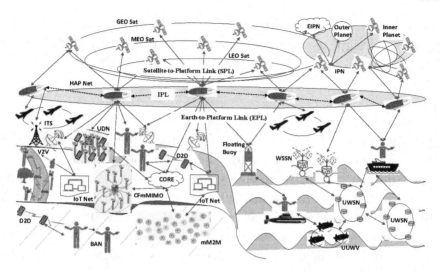

FIGURE 8.1

Overall architecture for mixed-mode communication system; GEO - Geosynchronous Earth Orbit, MEO - Middle Earth Orbit, IPL - Inter-platform Link, IPN - Inter-planetary network, ITS - Intelligent Transportation System, V2V - Vehicle-to-vehicle, D2D - Device-to-device, CFmMIMO - Cell-free massive MIMO, mM2M - massive machine-to-machine, UDN - User-defined Network, WSSN - Water-surface Sensor Network, UWSN - Underwater sensor network, UUWV - Unmanned Underwater Vehicle, EIPN - Exo-IPN.

All types of communication systems exploit naturally occurring physical phenomena like movement (displacement, diffusion, etc.) of fields, waves, and particles through different media. The concept of harmonic oscillators is very popular among scientists for characterizing physical phenomena. They have been separately used in the classical and quantum worlds to represent phenomena related to movement of waves and particles. Yet they have never been used by engineers developing communication systems and networks to model information flow in such a system. We need to bridge the gap between engineers and physicists in ways they use to model naturally occurring physical phenomena and systems that use them to communicate information. We need to develop a tool (mathematical and simulation-based) where a system engineer can input parameters of any communication system they want to design, media of information transfer and modes of communication used; and he/she should be able to estimate the system design details.

One of the most popular models used to characterize signal propagation in classical communications is the Nakagami-m distribution which is used to study the attenuation of wireless signals travelling in a multipath environment. Now it is used in applications as diverse as medical imaging and reliability modeling in manufacturing. Similarly, the concepts that will result from developing a generalized model for information flow can be extended for use in other fields of study like physics, biology, hydrology for modeling flow of signal in different applications.

8.2.3   Outlook

We can start by formulating the fundamental two-way wave equations and solving them for different environments by using different techniques. The two-way wave equations can be used to characterize the interaction between the fundamental propagating field (wave) and its time-delayed and attenuated copies reflected by obstacles in the environment, when electromagnetic and acoustic waves are used for carrying information. The two-way wave equation can describe both electromagnetic and mechanical wave propagation in any complex medium. However, the way they are solved depends on the medium under consideration. For example, for underwater acoustic communication, discretization of variable-density acoustic wave equation to quantify the filed strength. For quantum-based communication, analogy between paraxial wave equation and stationary Schrodinger equation can be used to characterize the travelling quantum field. For molecular diffusion-based communication, analogy between one-way wave equation and one-way diffusion, advective diffusion and turbulent advective diffusion equations can be used to calculate the propagating diffusion field.

How a propagating field carrying information travels between any two points depends not only on the medium and carrier of information but also on the geometry and anomalies present in the environment between those two points. To model the geometry of the environment, we can incorporate the concept of Whitney triangulation. The idea is to divide the space between two communicating points into millions of small triangles. Of equal dimensions. The number and size of the triangles depend on the shape and dimension of the space (for example, dimension of a room, distance, terrain and gradient between base-station tower and a mobile station). Now the space between the communicating points can have different big and small objects attributing to effects like reflection, shadowing, diffraction, obstruction, absorption, and penetration loss.

Instead of characterizing these effects separately, we can introduce the concept of harmonic oscillator to model the interaction between the anomalies in the environment and the travelling field carrying the information. The idea is to represent each of the triangles with a harmonic oscillator and then solve for each oscillator depending on the characteristics of the space. The interaction between adjacent triangles in space can be described by coupled oscillators where the oscillators are connected in such a way that energy is transferred between them. The triangles and the corresponding oscillators at the boundary of the propagating field carrying information will be directly interacting with the travelling field. This interaction will be modeled by a parametric process, that is, the propagation field is modulated by the spring constants of the oscillators at the boundary. Solving the system of coupled harmonic oscillators and interaction between the oscillators and the propagating filed, we should be able to calculate the received field strength at any point in space.

Additive phenomena like noise and interference also affect the propagating field in a wireless communication scenario. Interference can be easily described by the partial superposition of multiple travelling waves through the medium. For modeling noise, we can however resort to already available knowledge in literature, particularly use empirical distributions traditionally employed to model noise. Environment-specific noise distributions can be used to extract their individual mean, variance, and spectral densities. Once we have the received signal strength and the noise spectral densities, it will be possible to calculate the received signal-to-noise ratio (SNR).

Bibliography

[1] T. Khan, M. Civas, O. Cetinkaya, N. A. Abbasi, and O. B. Akan, "Nanosensor networks for smart health care," in *Nanosensors for Smart Cities.* Elsevier, 2020, pp. 387–403.

[2] K. Yang, D. Bi, Y. Deng, R. Zhang, M. M. U. Rahman, N. A. Ali, M. A. Imran, J. M. Jornet, Q. H. Abbasi, and A. Alomainy, "A comprehensive survey on hybrid communication in context of molecular communication and terahertz communication for body-centric nanonetworks," *IEEE Transactions on Molecular, Biological and Multi-Scale Communications*, vol. 6, no. 2, pp. 107–133, 2020.

[3] S. Ghafoor, N. Boujnah, M. H. Rehmani, and A. Davy, "Mac protocols for terahertz communication: A comprehensive survey," *IEEE Communications Surveys Tutorials*, vol. 22, no. 4, pp. 2236–2282, 2020.

[4] I. F. Akyildiz, M. Pierobon, and S. Balasubramaniam, "Moving forward with molecular communication: From theory to human health applications [point of view]," *Proceedings of the IEEE*, vol. 107, no. 5, pp. 858–865, 2019.

[5] L. Luo, X. She, J. Cao, Y. Zhang, Y. Li, and P. X. Song, "Detection and prediction of ovulation from body temperature measured by an in-ear wearable thermometer," *IEEE Transactions on Biomedical Engineering*, vol. 67, no. 2, pp. 512–522, 2019.

[6] U. M. Marcinkowska, "Importance of daily sex hormone measurements within the menstrual cycle for fertility estimates in cyclical shifts studies," *Evolutionary Psychology*, vol. 18, no. 1, p. 1474704919897913, 2020.

Index

Printed in the United States
by Baker & Taylor Publisher Services